ENGINEERED
TO SPEAK

ENGINEERED TO SPEAK

Helping You Create and Deliver Engaging Technical Presentations

Alexa S. Chilcutt
Adam J. Brooks

IEEE PCS Proefessional Engineering Communication Series

WILEY

Published by John Wiley & Sons, Inc., Hoboken, New Jersey.
Published simultaneously in Canada.

For general information on our other products and services or for technical support, please contact our Customer Care Department within the United States at (800) 762-2974, outside the United States at (317) 572-3993 or fax (317) 572-4002.

Wiley also publishes its books in a variety of electronic formats. Some content that appears in print may not be available in electronic formats. For more information about Wiley products, visit our web site at www.wiley.com.

Library of Congress Cataloging-in-Publication Data is available.
hardback: 9781119474968

Set in 10/12pt Times by SPi Global, Pondicherry, India

Printed in the United States of America

V10013064_081519

To the STEM professionals who inspired this book and improve our world, we hope you embrace your ideas and enjoy great success.

From Alexa:
To my parents, Dr. Gary and Diane Stough, who taught me to say "Yes" to challenges that are daunting, then go about figuring out how to accomplish them. Thankfully, the evidence of this upbringing is also seen in my siblings, Boyd and Laurel, and my own children, Natalie and Gary. To my aunt Rhonda Wilson whose constant enthusiasm about my accomplishments is the most gracious act of love. And to my husband Nathan who acts as my personal cheering section, endures the trials of a challenge-driven partner, and whose positivity is priceless.

From Adam:
To my late grandfather Sal Arello Jr. who dreamed of one day being able to hold a book with my name on it and whose reminder to always "think before you speak" framed so much of who I am today. To Cassie Price and the Bradley University Speech Team for your mentorship and training which is reflected in my passion for communication. To my parents Paula and Keith Sharples and my siblings, nieces, and nephew for the laughs and love. And to my husband Chris for your unshakeable faith in what I can accomplish and your persistence that I always strive for greater things.

CONTENTS

PART III: COMMIT TO IMPROVEMENT

A NOTE FROM SERIES EDITOR

For modern engineers, technical knowledge is not enough. These professionals also need the skills to communicate their ideas effectively and create opportunities for collaboration, support, and career advancement. In fact, it is unlikely that an engineer will find success without cultivating these interpersonal or soft skills.

Soft skills can be squishy and intangible, and their presence in STEM curricula is secondary to subject-matter expertise. More important, students do not always see the value in learning soft skills, such as verbal and written communication [1]. These students are naturally focused on acquiring technical knowledge, and do not yet see themselves integrated within the strategy of a larger organization. However, our everyday use of text messaging and cellphones has decreased our live, interpersonal conversations, and the associated skills are decreasing as a result.

Companies like Bank of America and Subaru now offer soft-skills training to employees on topics, including shaking hands, making eye contact, showing up to work on time, and wearing appropriate attire [2]. Technical knowledge will always be valued but being *workforce ready* requires competence in both technical and soft skills. In particular, engineers often struggle with maximizing their speaking opportunities in ways that clarify their ideas and connect with a broader audience. This book addresses this gap.

Engineered to Speak tackles the topic of verbal communication in practical and applicable ways. Most existing communication texts offer high-level content on a variety of technical genres or try and cover both written and verbal communication in their entirety. But the goal of Alexa Chilcutt and Adam Brooks is simple: they want to make you better speakers, better communicators, and better leaders. They dispel the myth that engineers are inherently poor speakers and then build a new foundation for communication success.

Alexa and Adam coach you toward recognizing speaking opportunities in the workplace and through the process of brainstorming, preparing, collaborating, practicing, and presenting your technical information. You have no better guides than Alexa and Adam. Their extensive coaching experience coupled with insights from practicing engineers have produced a text that is tailored for twenty-first century technical talent.

On a personal note, this is the first title under my editorship for the series in *Professional Engineering Communication*. My goal is to deliver quality content that directly applies to your STEM career. Theory still has its place here, but that theory drives toward empowering you with results-focused communication skills. This is not a series influenced by ideology and activism, but one focused on evidence and pragmatism. I hope that future titles elevate your classroom learning experience or your firm's next communication workshop.

Every successful series has an amazing team of people behind it. I appreciate the support from the IEEE publications team led by Vaishali Damle and her fantastic staff. Thanks also to Mary Hatcher and Nicole Hanley at Wiley. I only have the time to edit this series because of the support of my academic institution. I have called the Department of Technical Communication at the University of North Texas my home for almost a decade. Many of my colleagues are also my friends, particularly my chair Kim Sydow-Campbell. These people all make me a better editor and a far more patient, thankful person.

RYAN K. BOETTGER, PHD

REFERENCES

1. Pulko, S.H. and Parikh, S. (2003). Teaching 'soft' skills to engineers. *Int. J. Electr. Eng. Educ.* 40 (4): 243–254.
2. King, K. (2018, December 18). *Wanted: employees who can shake hands, make small talk.* https://www.wsj.com/articles/wanted-experts-at-soft-skills-1544360400?mod=e2tw (accessed 09 January 2019).

ABOUT THE AUTHORS

Alexa S. Chilcutt, PhD, is an expert in the areas of public speaking, leadership, and team communication. With a professional background in public relations, she has been published in *Public Relations Journal, Journal of the American Dental Association, MedEdPORTAL*, and various state and national trade magazines on topics surrounding organizational branding and team leadership.

In 2018, Alexa was awarded The University of Alabama's Knox Hagood Faculty Award for the College of Communication and Information Sciences where she also serves as Director of the Public Speaking Program and Communication Instructor for the university's Aeronautical and Mechanical Engineering REU Site: "Fluid Mechanics with Analysis using Computations and Experiments" (FM ACE), funded by the National Science Foundation. Alexa develops communication curriculum for various disciplines and is a regular provider of professional development training for corporate teams.

Adam J. Brooks, PhD, is a communication expert and a two-time national champion public speaker focused on helping people make their ideas more effective through the power of speech. With a PhD in Communication and Information Sciences, MA in communication studies, and a background in Public Relations, Adam has coached numerous individuals to national acclaim. As the Director of The University of Alabama's *Speaking Studio*, Adam is responsible for a campus-wide communication center focused on the training and development of presentation skills for young professionals.

Adam's work has been recognized by the National Communication Association, The National Speech and Debate Association, Regions Bank, The Community College Fellowship Program, and The National Forensic Association as well as college wide awards for outstanding teaching. As an assistant professor in communication studies, Adam is an expert in the areas of public speaking, strategic communication, rhetoric, and delivery.

Together, Alexa and Adam are the creators and instructors of the training workshop, *Professionally Speaking*, offered by the university's College of Continuing Studies. External to the university, they provide training to corporations around the world. Their method of training and ongoing support for professionals seeks to design profitable solutions to those delivering high-stakes presentations and pitches.

ACKNOWLEDGMENTS

A book is a tough task to undertake. In the beginning, we engaged in worrisome visions of traversing a difficult path alone. The reality was that creating this book was more about collaborating and creating community than we ever realized. In that same spirit, we want to acknowledge the wise, generous, funny, and motivating individuals who guided and propelled us along the way.

We are grateful to the army of practitioners who contributed their experience and insight to make this work relevant to others. Special thanks to Diane Sherman for creating her Sphere of Operation model one afternoon when reflecting on a previous conversation about the book's purpose. This model evolved into the Sphere of Influence that has guided much of our thinking.

To those who took the time to speak with us about the role of oral communication in their career, we are indebted to you. Without your personal narratives and examples, this would be like any other book. Thank you to – James Hans at Mercedes Benz US International; Daniel Schumacher at Torch Technologies, Inc.; Jenn Gustetic at NASA; Yuri Malishenko, Agile Coach at Danske Bank; David Daughton at Lake Shore Cryotronics; Noah Zandan at Quantified Communications; Steve Butsch at PowWow Mobile; Rajesh Mishra at Caterpillar Inc.; Tera Tubbs at City of Tuscaloosa, AL; Boyd Stough at Espy Revenue; Chris Ceroici, PhD Candidate at the University of Alberta; Sushruta Surappa, Graduate Research Assistant at Georgia Institute of Technology; Alexander Matlock and EcoCar Team 3 at The University of Alabama; and Annie Kary, a participant in UA's aeronautical and mechanical REU program and undergraduate student at Smith College.

This book would not have become a reality without the support, encouragement, technical expertise, and voice of reason of our Wiley-IEEE editor Ryan Boettger. Ryan took on more than he bargained for when he partnered with us, but we can never repay him for extending this opportunity and helping us develop our voice as writers to make the book readable and enjoyable for our audience. Thank you to Destini Smith for her graphic art work, and the entire team at Wiley-IEEE for bringing this project into fruition.

A project like this requires significant time to research and develop. As such we must recognize our incredibly supportive and encouraging Dean, Dr. Mark Nelson, and Department Chair, Dr. Beth Bennett at The University of Alabama's College of Communication and Information Sciences who were never short on wisdom and reassurance. Thanks also to our supportive colleagues in the Department of Communication Studies.

There are specific individuals whose advice has long shaped our thinking, and without whom this opportunity would not have been possible. To Dr. Kim Campbell at the University of North Texas for connecting Alexa with our university's aeronautical and mechanical engineering REU program in 2011 and more recently to Ryan Boettger at Wiley-IEEE, and to Brenda Truelove at UA's College of Continuing Studies who was one of the first people to invest in us and help realize the potential of our expertise through the development of the Professionally Speaking workshops.

Finally, we would like to thank our former and current students and graduate teaching assistants in our Public Speaking Program. These young professionals have taught us as much about the value of speaking skills and the most practical parts of education as we purport to teach them. Special acknowledgements go to the STEM students we have taken under our wing over the years and had the pleasure of witnessing their development as confident speakers and presenters. Their influence can be felt in each chapter of this book! We look forward to future trips on their yachts.

PART I

RECOGNIZE COMMUNICATION OPPORTUNITIES

1

WHY *THIS* BOOK?

If you cannot – in the long run – tell everyone what you have been doing, what you're doing has been worthless.
—Erwin Schrodinger, Winner of the Nobel Prize for Physics [1, p. 7–8].

*   *   *

The scene is familiar: a packed conference room with bad florescent lighting, the same staple of Danishes, slightly burnt coffee, and mini packets of creamer filling tables as attendees file in for yet another presentation.

At the end of a three-day deluge of PowerPoints and stale air, two team leads have been asked to give project updates at a meeting organized by a major manufacturer.

One team puts together a presentation they think is clear and logical. Since their audience also funded the research, the team plans to detail exactly how they spent their time. Slide-by-slide, they report every technical specification and graph consulted. As they progress through the slide deck, the team sees the audience's eyes glaze over, hears occasional sighs, and watches bodies slump with boredom.

The second team makes a different choice. They organize the material based on what the audience will find most interesting, omitting a portion of the technical details to be addressed during the question-and-answer period. The spokesperson for

Engineered to Speak: Helping You Create and Deliver Engaging Technical Presentations, First Edition.
Alexa S. Chilcutt and Adam J. Brooks.
© 2019 by The Institute of Electrical and Electronics Engineers, Inc. Published 2019 by John Wiley & Sons, Inc.

the team opens with only a few words on a slide and adds, "I will not bore you with details that you can read in the proposals and packets in front of you. Today, I want to share what was most interesting with our test and how our results can solve your problem." The presentation ends with time to spare and questions from eager audience members.

In conference rooms and online meetings around the world, engineers are reading their presentations directly from text-heavy slides, cramming too much information into 20-minutes, and leaving audiences without a clear understanding of what is important. You have sat through your share of mind-numbing project reports, and, if you are honest, you have probably delivered them too.

In your experience you have probably asked yourself, why are some ideas chosen over others? Why are some projects given funding, while others grow stale? Why are some individuals hired or promoted when others are not? The answer to these questions comes down to communication.

The ability to present information in a clear and compelling manner conveys expertise, enlists support, and makes favorable impressions. The engineering profession is an "intensely oral culture" [2, p. 12] that includes a variety of public speaking occasions, such as team interactions, communication with management, meetings, project updates, and formal presentations [3]. We have worked with hundreds of engineers at various stages of their careers. This book is the outgrowth of years of research, practice, workshops, and symposiums. Through our roles at a major university, we have perfected the ability to train others how to clarify and craft their messages, cultivate dynamic delivery, and calm public speaking anxiety.

A great deal is at stake when we step up to speak. The ways in which we make our expertise known and ideas accessible is at the heart of this book. Our goal is to answer one essential question for engineers and technical professionals: how can I get better at sharing my ideas with a variety of audiences?

We want to make you better speakers, better communicators, and better leaders. This text was not written to be a compendium of thousands of years of communication research, or as an academic text on how communication is relevant to engineers. Rather, we aim to do what engineers love to do most: break something down and figure out how it works. To achieve this, we combined our expertise in communication with current research and firsthand information gathered through interviews with practicing engineers and technical professionals around the world.

By reading and working through this book, you will learn how to

- make the complex simple and the simple interesting,
- craft clear and organized messages,
- speak to what matters most to any audience whether technical or nontechnical,
- stay focused on the desired outcome/goal of your presentation,
- design effective visual aids that work to enhance spoken messages, and
- manage anxiety associated with speaking occasions.

1.1 WHY NOW?

People associate engineering with intelligence, technical skill, tools, machinery, and complex design and drafting calculations. However, left out of this definition is the human element involved in helping to make an engineering project successful. The top 10 countries who provide engineering education currently produce 1,831,699 graduates annually [4]. For those in the STEM fields, technical skill and knowledge is not enough to succeed. The ability to communicate ideas effectively to technical and nontechnical audiences is vital to a project's support and career advancement. Tracy Robar found that "engineers who communicate well stand out from others in their field and generally have more success in engineering pursuits, while those who communicate poorly often find themselves unable to advance, no matter their technical expertise" [5, p. 26].

Further, a 2003 study surveyed practicing and retired mechanical engineers about their oral communication practices and frequencies in the workplace [2]. Fifty percent of these participants named some form of public speaking as crucial to their work. Thirty two percent named meetings as the most important public speaking opportunity. While the types of oral communication varied, 70% of respondents identified career advancement as a result of one's ability to communicate effectively. In other words, engineers are required to verbally interact with technical and nontechnical experts in order to advance their professional careers.

Engineering and communication are not rivals, they are partners depending on one another for success. Engineering combines both argument and science. The English word for technology comes from the Greek words *techne*, or art and *logos*, meaning speech or principle of logic. The early Greek philosopher and godfather of communication Aristotle used *logos* to mean argument. Here technology might be better thought of as the craft of logical argument. By extension, the term *engineering* might be best conceived as *technical management* [6]. Moving beyond ancient history, the field of engineering has negotiated the tension between useful practicality and the need to communicate its ideas. We tend to think of engineering as the systemic design and development of products [7]. In *Think Like an Engineer*, author Michael Davis details the history and ethics associated with engineering. He writes that engineering is the knowledge of how people and tools work together [6]. While your own education and experience have prepared you for ingenuity in the field, most engineers woefully neglect the people aspect of their profession.

Think for a moment about your day-to-day interactions. You have, no doubt, sat through presentations where a speaker crams every data point into a slide, or have a colleague who takes 10 minutes to get to the information that matters most. Maybe you have delivered a pitch or presentation that resulted in a less than enthusiastic response and missed opportunities for collaboration, funding, support, or buy-in.

The common thread is communication.

We believe that effective communication is attainable to any engineer or technical professional. While it may seem complicated, and damn near magical sometimes, this book is meant to break down the parts of being a better communicator and give you a step-by-step action plan for your next presentation or meeting. Just as CAD software enables structural engineers to visualize their ideas in 2D or 3D, this book is designed to model communication skills that create richer opportunities for professional development. We take the complex and make it simple, the simple and make it interesting. We will show you how to do the same.

1.2 MANAGERS

Would you love to manage a team of well-rounded and workforce-ready individuals? While most engineering degree programs require some element of communication training in the curriculum, few provide the time or true expertise to build the necessary skills. As a result, many engineers are not equipped to communicate their ideas or to maximize the variety of speaking opportunities they encounter. This is a resource for workforce readiness, professional development, and a helpful tool for leaders to use when investing in their team's communication abilities.

With advances in technology, the need to recruit and develop technical talent is imperative. Let us help develop that talent in a way that saves time and reduces costs. Teaching employees how to synthesize ideas, take them down the hall, up to management, or out to broader audiences will increase your organization's value. We walk readers through the *why* and *how* of communication, then through the active process of skill development. This is not a simple how-to book. Included throughout the book are assessments, practical activities, and feedback materials that provide measurable outcomes and help track improvement.

In working with both academic engineering programs and corporations, we saw a need and developed a unique training curriculum that can be implemented in part or in whole and is included in the book as a supplemental resource. Clout is communicated, not assumed.

1.3 ENGINEERS AND TECHNICAL PROFESSIONALS

Communication opportunities influence your day-to-day experience. Whether you have an upcoming presentation directed toward a non-technical audience, a team meeting where you are providing project updates, or a conversation with a colleague, this book will teach you how to create effective messages. In Chapter 2, we address the common stereotypes of engineers and debunk the myths to what makes them successful speakers. The goal for you is to accept that improving these skills is completely within your realm of control. The processes offered within the text provide you with a repeatable plan for each type of communication event. Eventually these steps will become learned and more intuitive, making you a more competent and confident communicator.

1.4 STUDENTS

No doubt, your STEM program is competitive. Think about your classmates. Visualize the last class you were sitting in. Who was sitting to your right? Who was sitting to your left? You are seeing a tiny slice of your competition for internships, co-ops, and full-time jobs. Employers want technical expertise in addition to the ability to clearly communicate what you are working on, what you need from others, to collaborate, to report to management when necessary, and possibly provide formal presentations or project updates.

The risks associated with poor communication skills are real, and the benefits are tangible. In a job interview, you will be asked to talk about a project you have recently worked on. Could you succinctly speak about the project, its highlights, and your contributions in one to two minutes as you will learn to do in Chapter 6? Now, can you do it without verbal fillers (*uh, um, ah*), awkward pauses, or going off topic? It is harder than you think. How many professors or classmates have made you sit through overly long and unnecessarily complicated presentations? It does not have to be that way. You can be better. Let us help you gain a competitive edge by teaching you how to recognize the importance of good communication and help you begin to improve your professional skill set.

To enhance your education, we provide the methods to satisfy ABET's criteria for Accrediting Engineering Technology Programs. Student outcomes include, "Criterion 3: f. an ability to apply written, oral, and graphical communication in both technical and non-technical environments; and an ability to identify and use appropriate technical literature" [8]. To achieve this, all the concepts, assessments, and exercises within this book are aimed at building oral and visual communication skills.

Finally, as you are currently in the classroom, we are bringing the voices of experts around the world to you. From software designers, engineers, to an international agile coach and visual practitioner, you will read what they have to say about communication and its role in their work, past failures, and successes. This may be a book that you are assigned to read, but we hope it will be one you enjoy and use throughout your career.

1.5 EMBRACING YOUR POWER AS A PRESENTER

Here's the cold, hard truth: the keys to effective communication are extraordinarily simple. In fact, when we present these ideas to individuals, they are dumbfounded as to how they did not think of them all along. The fear of public speaking is not about the fear of failure, but the fear of success.

Why does it take a book like this to get you to realize simple, basic, and effective means of getting the right message to the right people at the right time? How have we, as a society, created a culture where verbal fillers, dry information, and poor visual aids are the accepted norm? If the tools to engaging audiences, furthering your ideas, and getting the right people on board to achieve your goals are so simple, why do millions of people express fear when they get up to speak?

People are not afraid to be bad communicators because then they are just like everyone else. From the conference table to the cocktail party, people choose to be bad at the basics. Our culture has conditioned us to see effective communication techniques as a lack of authenticity. From our political sphere to the slickness of salespeople, we often perceive those that are polished as being out of touch or insincere. As such, you condition yourself to avoid the simple steps that would otherwise allow you to succeed. People avoid preparing a powerful presentation because they choose to blend with their peers instead of standing out. We think of poor communication as acceptable, and we then perceive powerful presentation skills as beyond the obtainable boundaries. In a culture in which everyone blends together, the standards for sharing and creating messages bends toward mediocrity.

The secret to overcoming a fear of public speaking starts with coming to terms with one essential truth: you must embrace being exceptional. Becoming exceptional means becoming comfortable with your own power. Effective communication has always been associated with creating power. As far back as ancient Greece, the great thinkers understood that teaching public speaking was a path toward leadership. The first public education program in the world began with teaching people how to construct and deliver speeches because the ancient Greeks realized the ability to stand up and influence people to accept ideas and act is the path toward meaningful change.

Your personal and professional hang-ups with speaking may not be rooted in failure, but success. If you have control over how your peers and customers perceive you – and can intentionally drive the right message to the right people at the right time – then you are responsible for the outcomes. The first step in advancing your professional goals is never be afraid to stand out, to impress people with your abilities. Do not be afraid to make a sustained and long-lasting impression. As you apply the knowledge we give, do not be afraid if you stand out from your peers. Embrace it and watch your career take flight. In short, do not be afraid to be good at this.

1.6 HOW TO USE THIS BOOK

We want this experience to be active. In each chapter, you will find objectives to guide your learning, activities and assessments to complete, and calls to action that extend the ideas and practices to build confidence and establish competence. Use it as a learning tool and then as a resource for continued professional development.

To simplify the process, the book is broken into three parts. Part 1 focuses on learning to recognize communication opportunities through self-assessment and awareness in Chapter 2, mapping your own Sphere of Influence in Chapter 3, and becoming aware of situational contexts so that you can appeal to any audience in Chapter 4. In Part 2, we identify the methods for putting these principles into practice. We will show you how to organize a pitch or presentation in Chapters 5 and 6, how to design visual aids that compliment your message in Chapter 7, and then cultivate charisma in Chapter 8. Finally, in Part 3, you commit by learning how to practice, give, and receive constructive feedback in Chapter 9, amplify your message in Chapter 10, and use the supplemental tools for training and development.

1.7 CALLS TO ACTION

In reading the answer to "Why This Book?," hopefully you are ready to challenge yourself to take the ideas presented in this text seriously. In other words, if we ask you to answer questions or complete an exercise, do it. If you engage in putting our concepts into practice, we are confident you will begin to see concrete results. There are plenty of poor to average to good communicators out there.

Why not be great?

Turn the page, and let us begin.

REFERENCES

1. Schrödinger, E. (1951). *Science and Humanism; Physics in Our Time*. Cambridge, UK: University Press.
2. Darling, A.L. and Dannels, D.P. (2003). Practicing engineers talk about the importance of talk: a report on the role of oral communication in the workplace. *Commun. Educ.* 52 (1): 1–16.
3. Dannels, D.P., Anson, C.M., Bullard, L., and Peretti, S. (2003). Challenges in learning communication skills in chemical engineering. *Commun. Educ.* 52 (1): 50–56.
4. Khushboo, S. (2018). Countries that produce the most engineers. *World Atlas*.
5. Robar, T. (1998). Communication and career advancement. *J. Manag. Eng.* 14 (2): 26.
6. Davis, M. (1998). *Thinking Like an Engineer: Studies in the Ethics of a Profession*. New York, USA: Oxford University Press.
7. Mills, H.D. (1999). The management of software engineering part I: principles of software engineering. *IBM Syst. J.* 38 (2/3): 289.
8. ABET (2018). Criteria for Accrediting Engineering Technology Programs, 2018–2019.

2

DEMYSTIFYING COMMUNICATION AND ENGINEERING

A first step to becoming a more effective communicator is to address the myths and misnomers that surround speaking well. In this chapter, you will learn how to

- identify the stereotypes associated with communication and engineering,
- dissect the myths surrounding public speaking, and
- evaluate your personal communication and speaking skills.

Engineers are known for solving unsolvable problems, advancing the technological revolution, and landing a human on the moon. Engineers excel at taking complex problems and creating effective solutions, but communication is simple, which is why it is incredibly difficult. As engineers, you know that fundamental to solving a problem is identifying its root cause. It makes no sense to develop a solution without fully understanding the problem. Similarly, before we can teach you to be better speakers and communicators, we need to start with the current state of your communication habits and awareness. We begin by engineering the self, dissecting the stereotypes associated with engineers and argue that every engineer possesses the foundations to be an excellent communicator.

Engineered to Speak: Helping You Create and Deliver Engaging Technical Presentations, First Edition.
Alexa S. Chilcutt and Adam J. Brooks.
© 2019 by The Institute of Electrical and Electronics Engineers, Inc. Published 2019 by John Wiley & Sons, Inc.

2.1 THE PLACE TO START

Think of all the adjectives you would use to describe a typical engineer. What words come to mind? Go ahead, try it. Put one minute on the clock, and in the space below write as many adjectives or words you can to describe *engineer*. We will wait.

———————————————   ———————————————

———————————————   ———————————————

———————————————   ———————————————

———————————————   ———————————————

———————————————   ———————————————

———————————————   ———————————————

Now, look back over your list and count the number of positive versus negative adjectives. If there are more negative adjectives than positive, you may have internalized these messages and need to shift your perception of what you and your career are capable of.

To do this, you must reject narratives about technical professionals not being naturally equipped to be good communicators. We asked 45 senior aeronautical and mechanical engineering students to perform this same exercise. Their adjectives included

- introverted,
- too smart to talk to normal people,
- terrible public speakers, and
- boring/overly technical.

Some of these students expanded on their initial list with the following statements:

The stereotype is that engineers are awkward and anti-social, but they are not.

Many engineers can communicate effectively with peers and other engineers but tend to not have anything to talk about with business-type people. Thus, the stereotype is born.

Introverted, quiet, rely very heavily on technical jargon to convey seemingly simple ideas when communicating with non-technical people.

They aren't very good at communication unless the audience is engineers or technically minded. The things we try to explain are difficult to understand!

Managers have high expectations of technical talent. Culturally, they expect them to have above average intelligence and the ability to create innovative solutions. Those expectations do not generally include that they also be an effective public speaker or compelling communicator.

> At a professional networking event for aspiring engineers this joke was told to us:
> How do you tell an extroverted engineer from a regular engineer?
> The extrovert stares at the other person's shoes when they are talking to someone.

A graduate student offered this as a response when we told them we help STEM professionals become better communicators. In fact, throughout the process of writing this book, as we talked with colleagues and friends outside the field of engineering, we encountered a similar response – good luck. When we hear this, we smile because we know how wrong they are.

2.2 REJECTING STEREOTYPES

In our experience, we find that technical professionals hold stereotypical beliefs about themselves and their ability to communicate. Every summer, we work with a group of students in a National Science Foundation-funded summer research experience. While working with these future aerospace and mechanical engineers, we tend to hear the same things about the value of communication skills:

> *These are soft skills. Learning public speaking skills won't matter to what I am doing.*

> *I'm a numbers person; get me back in the lab where I'm comfortable.*

The perceptions and paradigms we hold are powerful, and those beliefs impact our behaviors and attitudes. We know from research that people tend to match their performance to meet stereotypical expectations in a process known as stereotype threat [1]. For instance, when managers rely on stereotypical expectations, they ignore the potential and capabilities of the individual [2]. These threats impact education, including tests and classroom performance as well as the workplace development. Within the STEM fields, the degree to which a role model (say a professor) exhibits stereotypical beliefs of individuals who belong in STEM fields has a direct relationship on STEM participation [3]. Thus, we conform to stereotypes when we are confronted with people who hold those beliefs.

When we hold negative beliefs about what a profession is capable of, we further inhibit the success of those seeking to enter that profession. Examples of engineering stereotypes include a tendency toward social isolation and a focus on technology over people [4]. Research shows that the public perceives engineering as a self-oriented career without the need for community [5, 6]. Computer scientists in particular are overly stereotyped as socially awkward "nerds" who are obsessed with computers [3, 7, 8]. When researchers asked a group of women why they believed engineering was not a viable career option, they pointed to the stereotype that engineering was time intensive and incompatible with motherhood [6]. When presented with a list of STEM stereotypes, women reported feeling dissimilar and disconnected from the words they saw [3]. These misperceptions are harmful not only to the individuals who hold such beliefs, but for efforts to recruit and increase participation among underrepresented populations [9]. These beliefs inform what others expect from their technical talent and how profession- als see themselves.

Despite cultural beliefs in these stereotypes, there are strategies to confront and reject these notions for more positive outcomes. *Cognitive shifting* is the term for basically snapping yourself out of the incorrect beliefs you hold. The best way to reject negative self-impressions is to replace them with stronger positive impressions from the same stimuli. To combat the perception of engineering as an isolated pro- fession without the need for communication, it is instructive to remember that people associate engineers with large-scale problem solving.

Rather than maintain that engineers tend to be poor communicators, consider examples of the positive views the public holds for the profession. Recent campaigns from the National Academy of Engineers found that people strongly linked engineers with professionals who make a difference in the world [10–12]. We also know that engineering and STEM-related fields rank high in public credibility and are widely considered one of the most trustworthy professions [13–15]. In fact, the value of one's identity as an engineer is shown to be a stronger predictor of one's likelihood to stay in the profession [9]. By reframing the perception of engineering as something com- patible with communication skills, we believe you will be better able to shift yourself, and take the first step toward becoming the communicator you want to be.

2.3 DEBUNKING THE MYTHS

All the great speakers were bad speakers at first.

Ralph Waldo Emmerson [16, p. 78]

Based on our experience with working with students and industry clients, we have created a list of seven common myths related to speaking. These myths become the narratives people tell themselves about the skills associated with speaking.

2.3.1 Myth 1: Good Communicators Are Not Anxious

If the thought of preparing and delivering a message in front of a group makes you sweat, join the club. Anxiety associated with speaking is common, and up to 33% of individuals experience severe anxiety, categorizing this fear as a social disorder [17–19]. We get it; public speaking is not easy. We have been there. We have been the bad speakers who walked off stage and knew we blew it and have worked to become better speakers. Perfect speakers are a unicorn. The associated nervousness or anxiety cannot be completely ignored or eliminated, nor should it be.

In a recent workshop, we asked 65 participants if they experienced anxiety associated with public speaking. Three-fourths raised their hands, and it is amazing to watch an audience's reaction at this point. We then admit that despite years of speaking to large crowds, we often take five minutes of quiet somewhere (generally in a bathroom stall) to get in the right headspace before facing them. The point is, it is normal.

Even great speakers experience some level of apprehension prior to taking the podium. Overwhelming anxiety is debilitating, it can feel like a swarm of butterflies bouncing around your stomach, but the more you understand about the process and practice by placing yourself in low-risk speaking situations, the more manageable it becomes. In fact, some nervous energy is an asset. In Chapter 9, we offer strategies to minimize anxiety and refocus that energy to the topic. The goal is not to eliminate the butterflies but teach them to fly in formation.

2.3.2 Myth 2: Some People Are Naturally Great Speakers

When people witness a confident and compelling speaker, it is easy to assume that it comes naturally. The seemingly natural speakers are few and far between. We have found those with performative backgrounds (e.g. choir, theater, dance) more easily adapt to placing themselves in front of an audience and sharing their ideas. Even they, however, must learn the how-to steps of formulating clearly organized and memorable messages.

We spoke with Boyd, a 36-year-old entrepreneur, turned lawyer, and now successful business consultant. In his early twenties, he served as an Air Force Crew Chief for the first Special Operations Maintenance Group and worked on MC-130H Talon11 aircraft. Boyd is intelligent, articulate, and not too afraid of being the center of attention. At 6′4″, it is easy for him to command a room. He shared a story with us about a disastrous presentation he gave years ago to a group of Angel Investors.

> I was so passionate about it [the project] that I was all over the place. I was extolling the virtues of this idea and just exactly how groundbreaking and revolutionary it was. After it was over, I had a guy come up and say, "That sounds really interesting … but I have no idea what you're actually talking about." At which the 100+ people in the room began to laugh. By the way, before he'd said that, I was convinced I had knocked it out of the park.

Boyd went back to basics and practiced the tried and true principles of organizing information effectively.

You may be familiar with all the complexities of a project, but it needs to be digestible for your audience. There is a way to outline and present complex ideas to a variety of audiences that works every time. Chapter 5 guides you through the entire process from how to open with an attention-getting device to developing your three main points, how to link the main points together effectively, and how to close with a call to action. Boyd realized it was not about his knowledge or passion, but about how he packaged and framed his message. Since then, he has become a more prepared and thoughtful speaker. It is a process.

2.3.3 Myth 3: Winging it Works

Winging it typically results in overall ineffectiveness. Those who do not feel the need to prepare fall prey to common speaking pitfalls. Typical signs of winging it include

- abundant use of filler words (*uh*, *um*, *like*) [20, 21];
- ineffective use of time;
- not answering the *why* question that targets the audience's interests; and
- lack of clear organizational structure, leaving listeners unable to follow and retain information.

The technology company Samsung learned this the hard way when they hired director and producer Michael Bay to launch a new series of products in the home television market. Samsung invested millions of dollars in its design of a curved television with increased retinal displays. Bay's job was to link his brand of creating visually stunning films to the notion that the new curved television would create the ideal home experience. He had written a script to be placed on the teleprompter but had not familiarized himself with the actual material. He was planning to read the presentation. When the teleprompter failed, he became flustered. The moderator attempted to present him with questions to prompt him, and, in the video, you can hear Bay say, "I'll try and wing it." Less than a minute later, frustrated with the situation, he walked off stage.

The outcome of practice is more predictable than that of winging it. In Chapter 4, we give you a series of questions to help guide effective preparation.

2.3.4 Myth 4: Need to Be the "Sage on Stage"

As a technical professional, you seek solutions to problems and end up with many of the answers. However, when you take on the role of speaker, it can be tempting to think you have to be a *Sage on the Stage*. This is the belief that by placing yourself in the position of speaker, you must have all the answers. Two things occur from a Sage on the Stage performance. The first is that you place inordinate pressure on yourself to act as the supreme expert, and the second is that you will present in lecture mode. Neither is beneficial. When this occurs, you end up discussing every detail of the project and forgetting to connect with your audience. It is exhausting and no fun for the audience. People do not want to be impressed by you; they want to know why the information matters to them.

Think back to your least favorite professor (there may be more than one that comes to mind), the one who talked *at* the class. These professors strode into the classroom, pulled up their slides, and pontificated on a variety of concepts and methodologies relating to the topic of the day. They did not appear connected to the class or interested in your questions or comments. For all you know, this professor may have been rather insecure up there and this style was a type of defense mechanism. We know of faculty like this. They want to get in, give the information, and get out.

An audience can find a Sage on the Stage arrogant and distant. In Chapter 8, we will show you how to connect and engage with an audience on a more personal level. The goal is to share information, to create interest, and engagement.

2.3.5 Myth 5: Data Is Supreme

This may be the most pervasive myth for technical professionals. The tendency is to rely on numbers, showing *all* data involved in a project or solution. The thinking follows that the data *is* the presentation. Audiences desire enough information for them to know the following: (i) you are credible; (ii) the most important updates, findings, or outcomes; (iii) numbers as they relate to their specific interests; and (iv) what to do with the information. This will not necessitate every graph, chart, and data set that informed your conclusion.

If you want to see a statistician get it right, watch any of Hans Rosling's TED Talks. Hans was a Swedish academic and statistician who began his career as a physician. Later a Professor of International Health at Karolinska Institute, he founded the Gapminder Foundation and with his son, Ola Rosling, and developed the software system Trendalyzer that created graphics and animated data. At the time of writing this chapter, Hans' "The best stats you've ever seen" TED Talk had over 12 million views. While the data itself is astounding, his style of presenting was about connecting the data to the audience and to make them care.

Your goal may not be to rival Hans, but there are many ways to create a narrative around the data and highlight what is most important for audiences to remember. Organizing the message, framing it for the target audience, and understanding your end goal are essential to formulating engaging presentations. In Chapter 5, we outline the organizational process that helps to frame your material, and in Chapter 7, you learn how to create visual aids that bring the data to life.

2.3.6 Myth 6: Time – I Must Fill It

Time is a parameter, not a requirement. Wait, let that sink in.

You have 30 minutes to present. You have organized and created a strong 15–20-minute presentation that compresses the information and highlights the most important points. But you have more in-depth information that you *could* include.

Option 1: You practice and feel confident to deliver the presentation as is and leave time for questions and clarification (explaining additional rationale and data).

Option 2: You feel the need to reach the 30 minute-mark and add in a greater amount of detailed data.

There is always more information that could be included, but its inclusion might not keep the presentation focused or engaging. Better to present a clear, organized, and impactful presentation that takes 15–20 minutes and leave time for questions and clarification. Delivering too much or overly complex information results in an audience's inability to maintain focus and process information effectively, experiencing what researchers call cognitive backlog [22, 23].

The worst thing a speaker can do is to go over time. How do you feel about a speaker who does not use time effectively? When you are feeling the amount of information is greater than the time you are allotted, remember constraint breeds creativity.

2.3.7 Myth 7: Extroverts Make Better Speakers

Extroverts may exude a natural energy, but those with introverted personalities have the ability to deliver meaningful and thoughtful presentations. Dananjaya Hettiarachchi, the 2014 Toastmasters International World Champion, stated that introverts tended to be more empathetic whole extroverts projected more, but that extroverts could project with too much intensity [24].

We have seen our share of extraverts with a charismatic nature, who claim no fear of public speaking and therefore put little effort in preparation or rehearsing. It rarely works out well. The advantage introverts have is that they tend to be thoughtful about how the content is structured and are mindful to bring the audience with them in an authentic way.

2.4 CALLS TO ACTION

Now that we have debunked the most common myths, your professional development starts with taking stock of your own abilities. Complete the following assessments to gain an understanding of where you are and where you want to go.

2.4.1 Communication Assessment Questions

Results from the Communication Self-Assessment will help you acknowledge your strongest areas of communication and where you may struggle. For each of the below statements, assess your personal ability using the scale from 0 (no ability) to 7 (great) as they apply to your work environment or professional situations.

0 – No Ability   7 – Great!
          0-------1--------2-------3-------4-------5-------6-------7

1. Feeling confident to share your ideas verbally.   _____
2. Communicate verbally with those in your team.   _____
3. Communicate verbally to those in management or leadership positions.   _____
4. Share ideas in a formal group setting (i.e. meeting).   _____

5. Advocate for a specific action or point of view.　　　＿＿＿＿＿
6. Understanding other's motivations/perspective.　　　＿＿＿＿＿
7. Listening to gain understanding.　　　　＿＿＿＿＿
8. Reading someone's body language.　　　＿＿＿＿＿
9. Awareness of the nonverbal (body language) cues you display　　＿＿＿＿＿
   during an interaction.
10. Self-awareness of vocal variety (volume, pitch, inflection) used ＿＿＿＿＿
    during interactions.

Review your scores. The items you rated 5 or above can be considered communication strengths. Think of situations where these strengths work to your advantage. The items that you rated 4 or less may be considered communication weaknesses, areas you need to become aware of and strengthen. Think of situations in your work environment where this lack of skill presents a barrier. Accept the challenge as we progress through the book to look for content or exercises that will help you to build upon these skills. For example, if you scored low on item 7 (listening), read Chapter 3 about the barriers to effective listening and apply the four active listening methods that we describe.

2.4.2 Level of Anxiety in Public Speaking Situations

You have been asked to speak to a group of professional peers or to deliver a presentation to potential clients/investors. Rate the level of anxiety you experience prior to speaking?

0 – No sweat!　　　　　　　　　　　10 – I'd rather be mauled by a bear!
　　0-------1--------2-------3-------4-------5-------6-------7------8------9------10

If you placed yourself at the lower end of the scale, you are less anxious about speaking to a group. That is great. There are a great number of other elements of the speaking situation and ways to improve your speaking and presentation skills. If you placed yourself at the higher end of the scale, meaning you are more anxious or would rather be mauled by a bear than speak to a group, we can help! Not only will you walk through the methodology to make the presentation good, but Chapter 9 will provide you with methods to gain control over any speaking situation and tactics for reducing stress associated with public speaking.

Which part of the speaking/presenting process do you find most challenging?

❑ Crafting effective messages that resonate with my audience.
❑ Cultivating dynamic delivery (use of nonverbal gestures, voice, and movement).
❑ Creating a strong close or "ask."
❑ Calming anxiety.

Identifying which areas of message creation or delivery you find challenging will help you focus on certain aspects of improvement. However, we will be asking you at a certain point to ask others for feedback of your presentations or shorter pitches to gain an objective perspective. We do this regularly, subjecting ourselves to constructive feedback for continuous improvement. There have been times where we thought our message or structure was clear, but someone's feedback told us otherwise. Be open not just to learning, but to professional growth.

How likely are you to apply creative ideas and tactics to make a presentation engaging or memorable?

0 – Not Likely   10 – Highly Likely

0-------1--------2-------3-------4-------5-------6-------7------8------9-----10

We encourage creativity. Get out there! Have fun with it.

2.4.3 Putting Knowledge into Practice

Your Turn

As you carry about your work week or attend your next lecture, look for opportunities to put the knowledge from this chapter into practice.

- Observe examples of data overload in the next project update or presentation or a colleague who defies general stereotypes associated with engineering or technical professionals.
- Find the best speaker you work with and ask about the time they put into preparing their update or presentation and if they are ever anxious about presenting to a group of peers or stakeholders.

In Chapter 3, you will learn from a successful engineer about how to take your ideas down the hall to impact success as well as identify your own Sphere of Operational Influence.

REFERENCES

1. Eschenbach, E.A., Virnoche, M., Cashman, E.M., et al. (2014). Proven practices that can reduce stereotype threat in engineering education: a literature review. *2014 IEEE Frontiers in Education Conference (FIE) Proceedings* (22–25 October 2014). Madrid, Spain, pp. 1–9.
2. Stanton, R. (2017). Communicating with employees: resisting the stereotypes of generational cohorts in the workplace. *IEEE Trans. Prof. Commun.* 60 (3): 256–272.
3. Cheryan, S., Siy, J.O., Vichayapai, M. et al. (2011). Do female and male role models who embody STEM stereotypes hinder Women's anticipated success in STEM? *Soc. Psychol. Personal. Sci.* 2 (6): 656–664.
4. Barbercheck, M. (2001). *Mixed Messages: Men and Women in Advertisements in Science*. New York, NY: Rutledge.

5. Chambers, D.W. (1983). Stereotypic images of the scientist: the draw a scientist test. *Sci. Educ.* 67: 255–265.
6. Cheryan, S. (2012). Understanding the paradox in math-related fields: why do some gender gaps remain while others do not? *Sex Roles J. Res.* 66 (3–4): 184–190.
7. Margolis, J. and Fisher, A. (2002). *Unlocking the Clubhouse: Women in Computing.* Cambridge, MA: MIT Press.
8. Schott, G. and Selwyn, N. (2000). Examining the "male, antisocial" stereotype of high computer users. *J. Educ. Comput. Res.* 23 (3): 291–303.
9. Jones, B.D., Ruff, C., and Paretti, M.C. (2013). The impact of engineering identification and stereotypes on undergraduate women's achievement and persistence in engineering. *Soc. Psychol. Educ.* 16 (3): 471–493.
10. Belanger, A.L., Diekman, A.B., and Steinberg, M. (2017). Leveraging communal experiences in the curriculum: increasing interest in pursuing engineering by changing stereotypic expectations. *J. Appl. Soc. Psychol.* 47 (6): 305–319.
11. National Academy of Engineering (2008). *Changing the Conversation: Messages for Improving Public Understanding of Engineering.* National Academies Press.
12. National Academy of Engineering (2016). NAE grand challenges of engineering in the 21st century. http://www.engineeringchallenges.org/17849.aspx (accessed 14 May 2019).
13. Jennings, S., Guay Mcintyre, J., and Butler, S.E. (2015). What young adolescents think about engineering: immediate and longer lasting impressions of a video intervention. *J. Career Dev.* 42 (1): 3–18.
14. Camm, T.W. and Johnson, J.C. (2017). Managing engineering talent: unique challenges to optimize the best and brightest. *Min. Eng.* 8: 1–2.
15. Nosek, B.A., Smyth, F.L., Sriram, N. et al. (2009). National differences in gender-science stereotypes national sex differences in science predict and math achievement. *Proc. Natl. Acad. Sci. U. S. A.* 106 (26): 10593–10597.
16. Emerson, R.W. (1904). *The Complete Works of Ralph Waldo Emerson, with a Biographical Introduction and Notes by Edward Waldo Emerson.* Boston and New York: Houghton, Mifflin and Company.
17. Glassman, L.H., Forman, E.M., Herbert, J.D. et al. (2016). The effects of a brief acceptance-based behavioral treatment versus traditional cognitive-behavioral treatment for public speaking anxiety. *Behav. Modif.* 40 (5): 748–776.
18. Kothgassner, O.D., Felnhofer, A., Hlavacs, H. et al. (2016). Salivary cortisol and cardiovascular reactivity to a public speaking task in a virtual and real-life environment. *Comput. Hum. Behav.* 62: 124–135.
19. Slater, M., Pertaub, D.-P., Barker, C., and Clark, D.M. (2006). An experimental study on fear of public speaking using a virtual environment. *CyberPsychol. Behav.* 9 (5): 627–633.
20. Spieler, C. and Miltenberger, R. (2017). Using awareness training to decrease nervous habits during public speaking. *J. Appl. Behav. Anal.* 50 (1): 38–47.
21. Clark, H.H. and Fox Tree, J.E. (2002). Using uh and um in spontaneous speaking. *Cognition* 84 (1): 73–111.
22. Gallo, C. (2014). *Talk Like TED: The 9 Public Speaking Secrets of the World's Top Minds,* 1e. New York: St. Martin's Griffin.
23. Brownell, J. (2016). *Listening: Attitudes, Principles, and Skills,* 5e. London, New York: Routledge.
24. Feloni, R. (2016). A world champion public speaker says introverts can make better speakers. *Business Insider,* 2016. https://www.businessinsider.com/champion-public-speaker-says-introverts-can-make-better-speakers-2016-5 (accessed 14 May 2019).

3

RECOGNIZING COMMUNICATION OPPORTUNITIES

No idea has value if it sits on your desk.
It's up to you to get it off the desk and down the hall.
—Diane Barron Sherman, former Materials Engineer at Dow Chemical.

*   *   *

Technical expertise aside, if you cannot articulate your ideas by taking them down the hall to a colleague, into a meeting, or to a broader audience, your clout is inherently limited. If you can, however, you will have an advantage for advancing your career, and your company value will be greatly enhanced. This chapter addresses the interpersonal communication opportunities that impact your career. In this chapter, you will learn how to

- recognize how communication creates opportunities to increase influence and advance ideas,
- understand how oral communication creates collaborations and connections to your work, and
- learn active listening techniques for improved comprehension and message construction.

Engineered to Speak: Helping You Create and Deliver Engaging Technical Presentations, First Edition.
Alexa S. Chilcutt and Adam J. Brooks.
© 2019 by The Institute of Electrical and Electronics Engineers, Inc. Published 2019 by John Wiley & Sons, Inc.

Speaking creates influence. When you cannot articulate information effectively, you end up with two alternatives. The first is that you rely on someone else to explain your information and its relevance to others. The second is that you risk your ideas not being heard and you experience less than ideal partnerships or resources. The ability to communicate, understand, and listen to another's most pressing need maximizes interactions with teammates, colleagues, and leaders. The sharing of ideas increases collaborative efforts, helps to create buy-in, and works to ensure the advancement of your ideas.

This chapter presents what we call your Sphere of Influence. It begins at work, in your teams, and in the opportunities to interact with broader audiences. Chapter 2 proved that communication was integral to your professional development. Now we offer a model of communication to help you identify opportunities to understand the purpose of communication and learn how listening enhances day-to-day interactions. We offer the story of one engineer's career advancement to show how communication skills played a crucial role in her success and illustrate the value of taking your ideas down the hall – and beyond her comfort zone.

3.1 THE SPHERE OF INFLUENCE MODEL

Diane Sherman, a materials engineer, worked in Epoxy Resins Research and Development of Dow Chemical. During her career, she was appointed to the Dow Chemical Board of Scientists and represented the composite groups in Epoxy Resins R&D as well as other key elements of R&D related to analytics and engineering. When a leadership opportunity presented itself, Diane discovered that her ability to connect with scientists, engineers, and management distinguished her from her colleagues. A typical day for Diane included communicating with a variety of internal and external audiences; she collected information from stakeholders, connected ideas, and built upon resources.

Recalling the role communication played in her career development, Diane created a model called the engineer's Sphere of Influence. In our interviews with technical professionals, we kept coming back to this model and realized it was the best way to get you to understand how effective communication is the key to expanding your career. This model illustrates the different operational quadrants of one's career and life, taking them beyond their personal comfort level. As we explain each section, relate the different quadrants to your own professional experiences (Figure 3.1).

At the center of the model is the Core, where your skillset speaks for itself. Your Core incudes your training and technical expertise, intrinsic knowledge, and learned experience. Think also of your individual work space where you exhibit these skills simply by doing your job. The Core is your comfort zone, and to create change you are going to have to move beyond it.

Beyond the Core, there are two expanding spheres. The sphere closest to the center represents typical opportunities for daily interactions and communications. The second is an expanding sphere with no limits, representing the opportunities

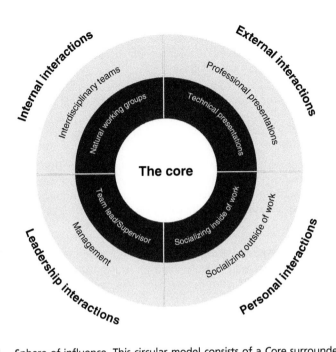

Figure 3.1. Sphere of influence. This circular model consists of a Core surrounded by three layered circles that are dissected into four quadrants and represent all possible communication opportunities for the technical professional. Source: Courtesy of Destini Smith.

made possible through communication to broader audiences. Additionally, the model is divided into four quadrants based on audience reach: Internal Interactions, Leadership Interactions, External Interactions, and Personal Interactions.

3.1.1 Internal Interactions Quadrant

The Internal Interactions quadrant is your most communal working group. It requires reaching beyond your autonomous work to interact with colleagues. Here, you interact with co-workers and working groups who communicate in the same technical language and share your same priorities. The outer layer stretches beyond your immediate workgroups to the challenge of interdisciplinary teams. These are teams where you may be assigned to a project or community effort and are communicating with colleagues who have varying technical skills, priorities, and perspectives. To illustrate this quadrant, read how Diane's ability to communicate across disciplines and groups helped connect the dots of the larger picture of research possibilities.

I became a supervisor of three different groups because I could speak about their research. I was the spokesperson for teams across disciplines from other departments for the same reason; I could advocate for their continued research, the importance of their work, and the impact to the company. I came to understand that my most valued strength wasn't my engineering skill–although I loved my work which was in fatigue crack propagation in composite materials and in building the materials science program. My most valued and recognized skill was the ability to communicate which resulted in my rapid advancement into leadership positions.

Let us begin to examine your existing opportunities of internal interactions.

Your Turn

Identify someone you work with on a project. Consider the shared language and understanding with which you operate. This is the reason you do not have to bridge the gap or reach to find a common language to update your co-workers and can speak casually using technical specific jargon.

Now, identify an interdisciplinary team member and list one shared and one different priority between this person and your natural working group. How would this impact your communication? We will discuss more on this in Chapter 4.

Interdisciplinary Team Member_____

Shared Priority/Differing Priority _____

Communication Impact _____

3.1.2 Leadership Interactions Quadrant

The Leadership Interactions Quadrant focuses on the ability to communicate up the channel of command. The inner layer requires communicating with direct supervisors. Here, you may interact with a team leader to discuss your own plans, ideas, or concerns. You may need to update someone on a project, request resources, or discuss goals and your progress against expectations. Most likely, this first layer of leadership shares common knowledge and language with you, although those in leadership positions have multiple or competing priorities. For example, you might need additional equipment and are requesting funding. Your supervisor, however, might have a list of other financial priorities competing with your request. While you may share much of the same technical knowledge or background, competing priorities necessitate a change in how you communicate specific needs.

When we spoke to a mechanical engineer with 10 years of experience, he discussed the distinction between management and engineers' concerns. This divide between competing motivations and knowledge is nothing new. Multiple studies of engineering team cultures demonstrate management is primarily concerned with costs and customer satisfaction, while engineers report being concerned most with safety and quality [1]. Understanding your counter-part's central interest and motivation will make you a more effective communicator.

The outer layer of Leadership Interactions moves beyond direct supervisors to upper management and senior leadership. Leaders at the top have a big picture focus, expectations that impact the bottom line, and vastly different knowledge and skills. Here, your communication must articulate how your work contributes to the larger organization, mission, or vision. Leaders at this level want solutions rather than problems. When you present a problem or concern, have an alternate plan to discuss. You need to verbally deliver messages in a clear and compelling way that helps to satisfy leadership objectives while obtaining your own objectives. Do not let your ideas sit on a desk. Keep in mind that if you do not stretch and adjust your communication skills in order to bridge this gap and take your ideas forward, someone else will.

Your Turn

Identify a team leader you communicate with on a regular basis. Also, identify someone who represents the highest level of leadership within the organization. Think about how you would frame the same topic for each of them and how you would approach them differently. If given the chance to speak about your work, what would you want them to walk away knowing?

Team Leader/Direct Supervisor_____

_____

Upper Management/CEO_____

_____

3.1.3 External Interactions Quadrant

The External Interactions quadrant crosses the boundary of the organization. In the inner layer, you have technical presentations or communications to peers who share your area of expertise. For example, you might interact with an engineer at another company, giving specs for a product.

Next, the outer layer involves presenting or speaking to groups who have a wide range of backgrounds, who may or may not share common knowledge. This includes client presentations, conference presentations, speaking to citizen groups, or pitching to potential stakeholders or government agencies.

As an example of both Leadership Interactions and External Interactions, one result of Diane's combined positions – a materials engineer and board member – was that she was frequently asked to speak to Dow Chemical's most important customers, top scientists, and upper management. As she spoke, she had to bridge across large gaps of shared knowledge and expertise.

Your Turn

Have you ever had to speak about a project to an external group? If so, who were they and what were the inherent challenges you faced? _____

_____

3.1.4 Personal Interactions Quadrant

Vital to quality of life, the Personal Interactions quadrant encapsulates speaking with ease about what you do and why it matters. The goal is to build community connection and personal relationships. Community begins with socializing inside of work and extends to outside of work to personal relationships in the community at large. Broadening your connections while remaining within the organization grounds the organization as a common reference point.

Moving externally to your community requires speaking about what you do and why it matters in a distilled and interesting way. Think of standing on the side lines of a football or soccer field where another parent asks what you do, or maybe you are at a social event or cocktail party. These are occasions to enhance your Sphere of Influence. You never know who you might meet and the impact they might have on your career. It is up to you to keep individuals who could have a stake in your work up-to-speed on progress and productivity, reminding them about how your work can impact theirs. Beyond the office, you may have the responsibility to speak with a variety of stakeholders, many of whom are nontechnical.

Your Turn

Identify an opportunity to connect with a person, a community, or organization. How might this connection benefit you personally or professionally?

_____

_____

3.2 COMMUNICATING TO CONNECT

No matter which quadrant you communicate in, there are two strategies for maximizing your Sphere of Influence: making connections and listening. Most people fail to think through a coherent strategy for how they engage with others. Because you have been talking since you were a small child, you do not feel it is always necessary to think through how you interact with others when trying to accomplish a crucial task. Part of this is biological. People underestimate the value of the things they do every day. We instinctively preserve our mental energies for complex tasks and devote little thought to everyday tasks like dressing or making our morning coffee.

These mental shortcuts in our daily processes are meant to preserve the brain's bandwidth. In *The Power of Habit*, author Charles Duhigg suggested that our neurology predisposes us to shorten complex processes into memorable and repeatable routines [2]. We are taught to recognize specific behaviors that trigger sets of automated routines and ultimately reward certain actions. In this *habit loop*, we reinforce activities that may be harmful to us because we are used to them and rewarded for repeating the behavior. While this may be useful in figuring out why you crave donuts at 3 p.m., it also explains why we rarely think strategically about our communication habits.

When your manager wants a project update, you could simply send piles of data, spreadsheets, memos, and scribbles of paper that were generated in the lab. We have all been in the meeting where an hour is spent receiving information that could have been transmitted via an email. Perhaps instead of writing this book, we could send an email with a list of sources you should read containing information points on how to speak well. However, after reading the list of pointers, you would be no closer to gaining the actual knowledge. Perhaps, then, we are not simply presenting to share information.

In fact, we have the opposite problem. We suffer from too much information, having access to more information now than at any other time. The proliferation of technology compounds this problem as every individual with access to a data connection and a smartphone has the capacity to publish and broadcast within seconds. Our ability to create and share information at the speed of our fingertips creates a perpetual cycle of noise. Social media only amplifies this problem. Every 60 seconds, over 3.3 million messages are produced on Facebook's platform, 65 000 posts are uploaded to Instagram, 400 000 tweets are sent, and over 500 hours are uploaded to YouTube [3–6].

When presented with an abundance of choice, we can feel overwhelmed and exhaust our abilities to choose the right information [7]. The concept of decision fatigue refers to the "phenomenon in which the limited reserve of stamina from decision-making becomes drained [with information overload] which leads to poor self-control" [8, p. 247]. Listeners also experience decision fatigue when someone delivers an avalanche of information without properly framing how it should be used or the desired action. In this marketplace of noise, we are also are speaking to make information meaningful and actionable to our audience. Chapter 4 covers audience analysis, and Chapter 6 details how to formulate a powerful pitch. For now, let us examine the impact of listening and its power to help inform us about what motivates others.

3.3 LISTENING HANG-UPS

The busy executive spends 80% of his time listening and doesn't hear half of what is being said.

—Harvard Business Review [9, p. 85]

People's time and attention is limited. Research shows that while the listening effectiveness rate is approximately 50%, after a communication interaction the listener loses one half to one third of what was said within eight hours of hearing the message [10]. This knowledge makes the case for understanding who you are talking to and drilling down to what they need to know and why. Listening is key to making your ideas instrumental to others. So how do we tap into our Sphere of Influence to discover what others want? How could your information fill a knowledge gap or existing need? It is about listening. Listening allows us to learn from others, articulate appropriate responses, and find common ground to build tangible outcomes. Think of it as intelligence gathering. To do this, we need to address the barriers of effective listening and active listening techniques.

Most people are not trained how to listen and there are many inherent barriers. Let us look at the seven most common barriers and put them in the context listening to another's message.

1. Lack of interest – Simply put, someone is droning on about a project you are not interested or invested in. You tune the message out and let your mind wander.

2. Distracting delivery – The speaker exhibits a distracting behavior (e.g. jingling change in their pocket, saying *um* or *like* every third word, rocking or pacing back and forth). You focus on the distraction rather than substance of the message.

3. External and internal noise – External noise refers to audiological stimulation external to the speaker's voice. Some examples include loud equipment, cell phones, or people talking. Internal noise refers to physiological conditions that capture your attention. Say you are hungry or have a headache, so you are thinking of the lunch you missed or the aspirin you need rather than the other message being conveying. Additionally, internal noise includes thoughts that are emotionally charged concerning the person or the topic and can include internal chatter, pre-programmed emotional responses to the subject that include your past experiences or opinions.

4. Ambushing – Instead of listening to another's message, you are busy formulating your point or argument and waiting in anticipation for the other to pause or take a breath, so you can interject your point of view.

5. Listening for facts – Just get to it already! You are listening for the key point that matters to you and not the entirety of the message.

6. Faking attention – You learned it in grade school. The teacher looks your way and you nod your head, make eye contact, smile, or verbally engage with a *right*, *yes*, or *okay*.

7. Thought speed – Our brains simply work faster than the rate of the typical speaker, therefore, our thoughts wander even while we are attempting to listen.

It is easy to recognize each of these barriers and recall, possibly today, experiencing them during a recent interaction. Listening, like speaking, is a skill. And like any other skill, it will only improve by learning a more effective process.

3.4 ACTIVE LISTENING TACTICS

Building rapport and meeting other's needs begins with being interested in what others want. Active listening means mentally and physically placing yourself in a receptive and responsive mode. This must be intentional. You begin by telling yourself at the beginning of an interaction, "I am going to listen to what this person is saying and figure out their perspective on the subject." It is a mental shift from an anticipated response mode. There are four active listening methods: paraphrasing, priming, expressing understanding, and the use of nonverbal body language.

3.4.1 Paraphrasing

Paraphrase is to intentionally listen for a portion of content that can be repeated back to the individual. The advantages of paraphrasing include rapport-building and an opportunity for clarification. When you paraphrase back a portion of another's message, rather than launching into your response, it assures the speaker you were listening. When they hear you repeat a portion of their message's content, they feel valued and heard. Paraphrasing a point that may need additional context or information provides you with greater accuracy in formulating a response. Paraphrasing is as simple as, "What your saying is that your data produced X type of results."

3.4.2 Priming

Priming means asking a question rather than immediately responding. This works to your advantage when a question will supply additional information. Paraphrasing and priming are easily combined. "What I hear you saying is that last Friday when your team tried to do X, you got a negative reading. Do you think this is an equipment error or human error?" This is key to formulating thoughtful questions and responses.

3.4.3 Expressing Understanding

Expressing understanding is generally verbal confirmation that you understand the intention or motivation behind the message. Paraphrasing and priming are about content; expressing understanding is about the emotions surrounding the content. "So, I know you are really frustrated about..." When you express understanding, you acknowledge the person not just the message.

3.4.4 Use of Nonverbal Body Language

Nonverbal communication includes facial and body gestures that signal engagement. This is intuitive to some degree but realize that the speaker is reading your nonverbal cues as interest or lack of interest. Are you looking at the person speaking or distracted by the e-mail that popped up on your phone? Examples of nonverbal cues that signal engagement include, head nods, eye contact, facial display, and leaning in or toward the individual.

Tip to Tune In: Nonverbally place yourself in an active listening mode by turning one ear slightly in the direction of the speaker. The speaker sees your ear and reads your nonverbal cue as evidence that you are listening to them. More important, this prompts you to listen rather than respond immediately and reminds you of employing other active listening methods. You will begin to think about a portion of their message that needs confirmation, clarification, or could benefit by additional information. This one tactic will make you a better listener, a more appropriate respondent, and less prone to misunderstanding.

Tuned Out Tip: Uh oh, they are out. When you see someone's feet, shoulders, or body beginning to face away from you or toward an exit (like a door), wrap it up. They are signaling, whether they are consciously aware of it or not, for the conversation to be over. Become self-aware of interaction cues to monitor conversational flow.

3.5 PUTTING IT INTO PRACTICE

In this final section of this chapter, we offer examples from Diane Sherman's experiences at Dow Chemical of taking it down the hall. As you read through these examples, recall examples from your own experiences where you did this; where you took your ideas forward, and what obstacles you overcame to do so. Consider what potential positive benefits might have resulted from different actions and what negative impacts have occurred.

EXAMPLE 1

Diane graduated with a BS in Mechanical Engineering and a MS in Metallurgical Engineering, when she was first hired, she was asked to develop the methodology for measuring the fatigue crack propagation of epoxy resins. As this was within her general field of study, she was completely confident and comfortable in delivering the desired outcomes. Soon after her work began, it was brought to her attention that she would be teamed with members from a different department. Analytical and Engineering Services (A&ES) was a newly formed, relatively high-tech department, whose goal was to bring advanced analysis to the product-oriented departments, such as Resins R&D. Specifically to this example, they were to model the Fatigue Crack Propagation results that Diane had generated experimentally. Overseeing this collaborative effort was a senior scientist in A&ES who had a commanding and intimidating presence.

A major issue arose when the model developed by A&ES did not quite predict the results obtained by Diane's test methods. Diane faced a problem of credibility. She was newly hired and from a different department – not specifically regarded for its analytical science but had reproducible and more accurate results than the other team. She found herself faced with the decision to either step forward or to allow the A&ES team to take over the generation of the Fatigue Crack Propagation results using their model. The most critical discovery was when Diane understood that A&ES had a different (and high profile) set of priorities than her home department. Yet, she recognized that her success depended on the success of the joint project, and that she would have to find a solution which would meet both sets of priorities. Formulating a solution which addressed the priorities gave Diane the confidence to take this idea forward – to take it down the hall ... and across the street to a different building and into an office of a senior and intimidating scientist.

EXAMPLE 2

After moving into a supervisory role, Diane was given responsibility of the materials characterization lab for epoxy resins and composite materials. Feeling totally in her comfort zone, she scrutinized the various test methods and discovered a significant procedural error. The tensile testing method being used for composite materials was the same one that was being used for neat resins. Realizing the error and overall negative impact on the tensile properties of the composite materials, Diane immediately pulled all composites from the testing cue and worked to replace the testing procedures. Significant down time occurred to allow for machining new test fixtures and dies and the training of technicians. Meanwhile, every supervisor over the various epoxy resin product groups, learned through *their technicians*, that testing had stopped. They were furious. In addition to not knowing anything about this in advance, these supervisors had their own customers to answer to. Even though the reasoning was sound, the lack of communication on Diane's end jeopardized the transition to the correct procedure, and more importantly, her working relationships. Looking back, she realized how taking the idea down the hall and into each supervisor's office would have hastened the process. In fact, it was a lost opportunity because there was a good story to tell, in that the correct test method would significantly raise the measured tensile properties (improving the performance of their products).

From these examples, the first lesson is the importance of knowing the priorities of those you work with and shaping your message to address those concerns. The second lesson shows the importance of keeping people informed along the way, instead of after the work is completed.

3.6 CALLS TO ACTION

Take time to create your own Sphere of Influence using the model below. In each quadrant, identify opportunities to communicate with individuals or groups according to the sector. This activity will allow you to recognize potential areas of professional communication, growth, and development.

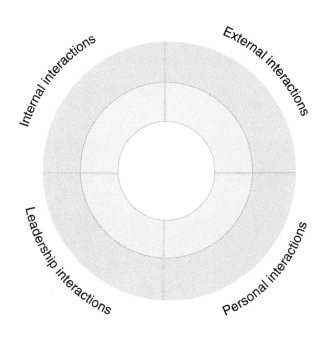

REFERENCES

1. Davis, M. and Association for Practice and Professional Ethics (1998). *Thinking Like an Engineer: Studies in the Ethics of a Profession*. USA: Oxford University Press.
2. Duhigg, C. (2012). *The Power of Habit : Why We Do What We Do in Life and Business*. New York: Random House.
3. Smart Insights (2017). What happens online in 60 seconds? [Infographic]. *Robert Allen Full*. https://www.smartinsights.com/internet-marketing-statistics/happens-online-60-seconds (accessed 01 August 2018).
4. Company Info|Facebook Newsroom (2017). https://newsroom.fb.com/company-info (accessed 26 October 2018).
5. Instagram – Info Center (2017). https://instagram-press.com (accessed 26 October 2018).

6. 1 Second – Internet Live Stats (2018). http://www.internetlivestats.com/one-second (accessed 26 October 2018).

7. Pocheptsova, A., Amir, O., Dhar, R., and Baumeister, R.F. (2009). Deciding without resources: resource depletion and choice in context. *J. Mark. Res.* 46 (3): 344–355.

8. Polman, E. and Vohs, K.D. (2016). Decision fatigue, choosing for others, and self-construal. *Soc. Psychol. Personal. Sci.* 7 (5): 471–478.

9. Nichols, R.G. and Stevens, L.A. (1957). Listening to people. *Harv. Bus. Rev.* 35 (5): 85–93.

10. Harris, T. and Sherblom, J. (2010). *Small Group Communication*, 5e. Pearson.

4

ASKING THE QUESTIONS

When preparing a presentation, do you open a blank PowerPoint and fill it slide-by-slide with text and data? If so, the results may be information overload and missed opportunities for audience connection. Most often, the typical approach is to allow information rather than situation drive the message.

A project begins with a defined objective. Preparing for a presentation is no different. Fully formulating a description of the speaking occasion takes time. In this chapter, we provide a systematic method of intelligence gathering to allow your presentation to achieve its desired outcome. In this chapter, you will learn how to

- identify the speaking occasion and its purpose;
- analyze different aspects of your audience, whether technical or nontechnical;
- understand constraints to message reception; and
- recognize whether your audience is tuned in or tuned out.

The questions Who? What? When? Where? Why? and How? have been used by speakers for thousands of years and are still the staple questions of journalists, researchers, and forensic investigators [1]. The common, but incorrect, approach is to take all your information and fill as many slides with text, charts, and graphs as necessary. But taking the time to answer these questions will help you identify the most relevant information and decide how best to present it.

Engineered to Speak: Helping You Create and Deliver Engaging Technical Presentations, First Edition.
Alexa S. Chilcutt and Adam J. Brooks.
© 2019 by The Institute of Electrical and Electronics Engineers, Inc. Published 2019 by John Wiley & Sons, Inc.

4.1 ASKING THE QUESTIONS

When you are engaged in a project, you begin with the goal in mind. Likewise, when you have recognized a communication opportunity, or are given the chance to speak in a meeting or to an audience, you need to ask yourself the following questions:

1. Who am I speaking to? Who needs the information?
2. What is the purpose of my presentation?
3. What is the desired outcome?
4. What information matters most?
5. Why should my audience care?
6. When am I speaking?
7. Where am I speaking?
8. How will I present? What types of types of visual aids would supplement my message?

Answering these questions allows you to consider the speaking occasion. Speaking is different than writing. When speaking, you have a limited amount of time to capitative an audience. If done well, you can gain their interest and engage with them while simultaneously informing or persuading them. The level of interest you create in the subject depends on how you construct the message. You must drill down to the purpose of delivering that specific information, discover as much as possible about your audience, and drive the message toward a specific outcome. To illustrate how these questions drive the information presented, let us consider at a real-life success story.

EXAMPLE

When Mercedes-Benz International expanded their operations, Jim Hans scouted a location for the new global logistics hub. Months of research, site visits in multiple states, logistical considerations, and number crunching led his team to recommend Alabama's impoverished Bibb County. As the Manager of Investment and Projects, Jim knew that presenting to key state politicians was critical to the success of the project. The stakes were high. Bibb County was convenient, just one county over from the current production location of the GLE and GLS SUVs and C-Class vehicles. How would he approach this audience and create buy-in that aligned with Mercedes-Benz's interests? Did he need to explain the data behind their rationale to recommend Bibb County over other sites?

Jim did not wake up the morning of the presentation, open a blank PowerPoint, and slap a slide deck together. The information he shared was strategically selected, logically presented, and artfully displayed. He began with considering the rhetorical situation, the needs and constraints of his audience, and how he wanted the audience to respond to his call to action.

Jim asked the following questions: What is my ultimate objective (purpose)? Who am I speaking to? Why is the information important to this specific audience? When am I presenting? Where? How much time do I have to convey the message? How should I use visual aids to present the message? With the answers in mind, he built the presentation around his objective: to persuade the Alabama Secretary of Commerce and Bibb County officials to partner with Mercedes-Benz. His approach was informed by tailoring the message around the "What's in it for me?" question for his audience. Jim's presentation was a success and a win–win for Mercedes-Benz U.S. International and a small Alabama town.

Like Jim, making the complex simple means drilling down to discover as much as possible about what motivates your audience. As we examine each of the questions, think of an upcoming presentation or pitch and answer each according to your specific situation. Let us look at the questions.

4.1.1 Who Am I Speaking to?

As we wrote in Chapter 3, your audience may work down the hall or they may be an external stakeholder with a nontechnical background. If speakers understand their audience, they can choose the most relevant information and present it in a manner that resonates. You will need to create a detailed profile of your audience – their degree of knowledge, interests, motivations, and predispositions toward the topic.

In Jim's case, he presented the team's recommendation for the site to the Alabama Secretary of Commerce and other state and local politicians. This message centered on creating jobs and improving infrastructure for the county while securing the necessary incentives for Mercedes. If he spoke to Mercedes executives instead, Jim would have emphasized the supply cost reduction benefits provided by the projects. By keeping one specific audience in mind, you have the ability to tailor your message to the people who need it most.

4.1.2 What Is the Purpose of My Presentation?

Are you informing or persuading? It depends on your objective. When informing, ask yourself if you are extending common knowledge or providing new information. These are straight-forward messages that create shared meaning and understanding. Building on the previous question, ask yourself what portion of the information your audience will be most interested in. If your objective is to persuade, consider how to create buy-in. Do you need the audience to adopt a certain viewpoint or take a specific action?

Jim's goal was to persuade – he needed to convince the Alabama Secretary of State and local politicians to build the logistics plant in Bibb County. Creating links between the information and the audience's interests are key to shifting attitudes and prompting action.

4.1.3 What Is the Desired Outcome?

You want to be specific in what you want the audience to do after you are finished speaking. For many, this might be as simple as wanting your audience to come speak with you after the event. You might also need the audience to invest their time or money (or both) into your project. Let your answer to this question guide how you format the rest of the presentation.

Jim said the best way to approach a presentation was to think of the outcome or action you want. "I start with the end in mind. My thought process is where do I want to get, what do I want to achieve? I'm in the role of a finance person but still think like an engineer. How do I create a step-by-step, flow process, knowing where I need be and thinking about how can I get there?" Preparing your presentation with the end in mind allows you to construct a clear path and help an audience reach the desired conclusion. Think of this process as reverse engineering the presentation.

4.1.4 What Information Matters Most?

Studies show that humans recall 17–20% of what they hear from an audio-visual presentation [2, 3]. In lectures, audiences recalled 20% of a 40-minute lecture, 25% of a 30-minute lecture, and 41% of a 15-minute lecture [4, 5]. If people are retaining 20–40% of what you say, make sure they remember the right content. What strategies can you adopt when an audience is absorbing half of what you are saying and only remembering a quarter of it? The most vital information is what you want the audience to remember or act upon. Organize the most relevant information under three main ideas – think *Beginning, Middle, End*.

4.1.5 Why Should they Care?

During any speaking occasion, the audience is asking, "what's it in for me?" How does your purpose align with their interests? Success comes from creating links between the message and the audience through information selection, logical and emotional appeals, and making sure they get the take-away or call to action.

In Jim's example, he knew presenting to a political audience required more than the standard slide deck of pure financials. In fact, his presentation did not include a single financial statement. The presentation opened with a slide show of photos that highlighted the 20-year relationship between MCBI and Alabama. Jim then moved to the problem his audience needed to solve: improving economic development in Bibb County. He built his entire presentation around communicating that benefit. He asked his audience:

> Don't you want to help us help you? We could go to Georgia among other states, but we think Bibb County is the right thing to do. We know it is on your "Four Troubled Counties" list, a place where everybody leaves to get employed. Hammering home the political aspect, I appealed to him [Alabama Secretary of Commerce] by saying "You're employed by the Governor, the Governor is elected by the people, here's how you deliver some promises to the people, but we need your help.

Jim recalled, "I laid out what a good thing it would be for the state of Alabama," and he presented Mercedes' wish list, which included "infrastructure, roads and

Highway 11 improvements, etc. making it better for us to do business and help employ 600 of their people." Jim's presentation was a success as the stakeholders announced the decision to build the 260-acre site in Bibb County as part of a one billion-dollar investment [6].

4.1.6 When Am I Speaking?

Time is a parameter, not a requirement. Creativity thrives within constraint, so look at limited time as a challenge to pique the audience's interest, deliver most relevant information, and leave them wanting more. The parameter of time lends a framework to organization. Now that you have listed the most important points, take the allotted time and reverse engineer the amount of information you can effectively cover.

When you attend events with guest speakers, you can determine how effective they will be based solely on the number of slides in the deck compared to the number of minutes allotted for that presentation. When you see presenters with 50 slides for a 45-minute talk, you should know you are in for trouble.

In a formal presentation, one strategy is to build in time for the audience to ask questions. Allowing time for questions permits you to respond with greater specificity and rationale. It also reengages the audience in the process.

4.1.7 Where Am I Speaking?

Location, location, location. While this question does not deal with content itself, it determines formality and delivery. Knowing the size of a group and layout of the physical space is beneficial to your preparation. If it is a larger group and space, a more formal tone and visual aids that can be seen and understood from a distance are advantageous. If it is a smaller, less formal group, an organized message with a conversational tone may better suit the situation. Spatial awareness works to your advantage in combatting public speaking anxiety as the ability to visualize delivering the message in the space decreases the ambiguity of the situation. If you cannot answer this question, prepare to be flexible and work with the existing space.

4.1.8 How Should I Present?

In a typical presentation you have two choices: verbal delivery or verbal delivery supplemented with visual aids (e.g. slides, demonstrations, physical objects, handouts). Engineers are often demonstrating how something works or discussing design; therefore, the use of visual aids is extremely beneficial. A meeting of peers may need a simple verbal update of information while more formal presentations with a wider audience or one that involves new concepts may need images or physical objects to reinforcement concepts. How you choose to present is directly tied to the purpose, audience, and time. Chapter 7 addresses how to craft effective visual aids and provide you with a variety of design techniques and examples.

4.2 ANALYZING YOUR AUDIENCE

The ability to adapt a message to a variety of audiences is key to overall communication proficiency. Your information is the foundation of any message, but the scope of detail, word choice, and delivery depends on your audience. Dr. Daniel Schumacher, the Senior Manager at Torch Technologies and former Director of Science and Technology for NASA, confirmed that engineers who are able to speak to a variety of audiences experience greater career advancement [7]. To prove his point, he spoke about working with recipients of the Bruno Rossi Prize in high energy astrophysics. As he networked, he observed their unique ability to switch between divergent styles of communication. According to Schumacher, the ideal technical professional is one who "can have an in-depth conversation with colleagues about black holes and turn around and explain what a black hole is to a group of eight-year-olds." In the following section, we will cover some considerations to make when analyzing any audience.

4.2.1 Captive Verses Voluntary Audiences

There is a difference between sitting through a presentation you are required to attend and electing to listen to someone because you have a genuine interest in them or their topic. Think of an upcoming presentation you are delivering.

Your Turn

Is your audience required to attend, or are they choosing to attend your presentation? _____

A captive audience is comprised of individuals who are required to be in attendance. A common example of this type is a typical meeting in an organization. In these instances, you are typically communicating what you or your team is doing and how it fits into the larger mission. Delivery with captive audiences can be more straightforward. Think of the meetings and presentations you are required to sit through. If everyone who spoke had a clear and concise message, used time efficiently, and made it matter to you, would it be so bad? Recognizing others are captive places pressure on the speaker to be prepared and organized. Want to turn an audience *off*? Waste their time. Your goal is to make it worth the time they have spent sitting through it.

Voluntary audiences are comprised of people who have an interest in the topic or the speaker. The voluntary audience wants to hear about you, your project, research, organization, or has a vested interest in the information being presented. Think of conferences where you choose a session depending on the topic or speaker. The voluntary audience can opt out at any point in the presentation. This may occur in the form of purposeful distractions, pulling out a cell phone or laptop, or simply walking out. Engagement is key. The speaker has an obligation in each scenario to provide and deliver the information according to the audience's need to know. In Chapter 5, we will cover organization that allows audiences to follow information easily.

4.2.2 Knowledge

Think again about an upcoming presentation. Depending on the topic, how much technical knowledge does your audience possess? If we go back to Jim's presentation, an example of measuring knowledge might be – how much does the audience know about the pressing economic needs of Bibb County? How much pre-existing knowledge of the subject does your audience have? Measure their awareness along a five-point Likert scale.

Little/None	Some	Unsure	Good Amount	Great Amount
1	2	3	4	5

On this scale, Jim's audience would score a 5. If the question were, "How much do they know about the incentives needed to make building the plant viable for Mercedes?" The answer could have varied greatly depending on each member. In this case, Jim would choose to spend a greater amount of time informing the audience of the rationale for incentives and less time trying to convince them of Bibb County's need.

Placing the audience's knowledge on a scale provides a visual cue that guides the amount of background information, speed of delivery, and use of technical language. If you are updating a group on a project, take a couple of minutes to review where you left off last time and then transition into new information. If your audience has little knowledge of the subject, build in time to explain or introduce concepts in a relatable way.

4.2.3 Technical and Nontechnical Audiences

Identifying your audience in this binary is a starting point for decisions about language use. Think back to your Sphere of Influence and the Internal Interaction quadrant from Chapter 3. It is easier to present information to a technical audience, especially one within your specific field, because you can use jargon and focus on moving the information forward. However, not all technical audiences are alike, make sure to account for the knowledge base of technical professionals outside your specific field.

With the non-technical audience, you must create shared meaning in the moment. This is one major difference between written and spoken communication. If your audience was reading your content and came across a term they did not know, they could take a break from reading to look it up. An audience listening to your message cannot take that break to bring themselves up to speed.

Your job is to explain a complex topic through simple terminology and vivid language that tells a story. A speaker must be cognizant of inundating an audience with too much or overly complex information or delivering it at a pace that leaves listeners behind. Think about framing the information based on what it can do for your audience. Later, we will discuss constraints associated with presenting to non-technical audiences, including the phenomenon of cognitive backlog.

4.2.4 The Jargon Barrier

Jargon refers to technical language used within a specific field or subset of experts. Among peers, this is a common language. So common, that it is easy to be completely unaware of the amount of jargon you casually use when communicating to laypeople.

For individuals outside of your knowledge base, jargon is often unintelligible and creates confusion or ambiguity. As a speaker, you must become highly attuned to the makeup of your audience. If the goal of communication is to create shared meaning, then the skill of simplifying and clarifying concepts is paramount to message reception.

Alexa's first post-doc position was to develop a communication curriculum for a Dental School. Her dissertation research had focused on dental leadership and team communication as she had grown up in a family of dentists and seen how communication skills played a major role in a practice's success [8].

The research landed her a position created in response to the American Dental Education Association's guidelines for Competencies for the New General Dentist, including Competency 3: Communication and Interpersonal Skills. She possessed more than enough knowledge and ability to succeed, but the environment of a dental school was completely unfamiliar to her. As the Dean of Academics spoke about his vision for the program, she understood about 60% of what he was saying and was too embarrassed to ask simple questions like, "What is a D2?" As a result, there was an immediate barrier to faculty engagement as she had to take the time to learn everyday dental school terms.

Non-technical audiences require you to think in terms of creating understanding without the use of highly technical terms. Tactics include the use of vivid language, narratives, metaphors, simplistic descriptive terms, or even broader generalizations that help connect the dots from your expertise to their interests. Simple illustrations, photographs, physical objects, and demonstrations bring abstract concepts to life.

The non-technical audience is anyone outside of an engineering or technical field, and to a degree other engineers or technicians outside of your area of expertise. Think of the individuals you placed within your External or Interpersonal Interaction quadrants in Chapter 3. These are people who may listen to your jargon and are unwilling to ask what you mean. How can you begin to think about broadening your Sphere of Influence by communicating more effectively with these nontechnical individuals?

4.3 COMPETING FOR ATTENTION

Your time is limited and so are people's attention spans. Jargon attributes to cognitive backlog as it presents complex language that creates uncertainty in the mind of the audience. This is compounded by an individual's ability to absorb and process information. For all audiences, cognitive processing while thinking, speaking, and listening is hard work. Extensive research from Harvard shows the average person listens to half of what is being said [9]. This immediate 50% effectiveness rate declines to 25%, two months after hearing the information [10, 11]. Speakers are not the only ones who experience anxiety in a communication interaction. Cognitive backlog occurs when listener's experience uncertainty. Uncertainty creates anxiety over lack of comprehension. Three factors contribute to ineffective message reception: thought–speed differential, complex information, or intake of too much information and not enough time to process along the way [9, 11].

Thought–speed differential takes place because our brains run at a much faster rate than the speed of someone's speech. The average person speaks at a rate of 150–160 words per minute and our brains are thinking at an average rate of 1000–3000 words per minute [12]. This differential results in day dreaming. When a listener tunes back into the conversation and realizes they have lost their place, rather than play catchup, they tune out. Complex information creates anxiety due to lack of comprehension. This feeling of misunderstanding places pressure on listeners and distances them from the speaker. Tuning out is the path of least resistance. You must be aware, as a speaker, of the audience's nonverbal feedback. Are they tuned in or tuned out? Are they looking at you with a quizzical expression? If so, stop. Ask questions to see if there is a need for clarification or move on to another point helps to re-engage the group. In Table 4.1, we list key body language indicators to watch for when gauging audience engagement.

T A B L E   4 . 1   Audience feedback indicators.

Tuned In	Tuned Out
Good eye contact	Body turned away
Head nods	No eye contact/Eyes shut
Smiling	Irritated/bored facial expressions
Leaning in or forward	Scrolling through phone
Writing notes and looking up at you	Reading another document
Ears turned toward you/trying to listen	Working (not note-taking) on computer

Signs of nonverbal communication indicating whether audience members are engaged or tuned out.

4.4 OPPOSING VIEWPOINTS

An audience is not a blank slate. They listen to a presentation primed with previous knowledge, experience, and currently held opinions. If you state an opposing viewpoint without taking your audience's position into consideration, you will experience resistance. It takes only eight seconds to lose someone's attention and when you present an opposing viewpoint, their internal argument of your point will divert their attention away from your message [10, 13]. In preparation, study what the audience knows about the topic. In your presentation, refer to their opinions or arguments before revealing why your recommendation suits their needs better than another alternative.

Considering the possible arguments grants you the opportunity to counter the audience's primed perspective. If your audience is perceived to be negatively primed to the topic, learn as much as you can about what is informing their position. Are they opposed to the topic or proposed action because of past experience, divergent values, or financial considerations? When you consider their points of view, you have a better chance of changing their minds. You will need to plan time to address their concerns.

The goal is to keep your audience positively engaged while steering them toward your call to action. Consider measuring your audience's favorability toward a topic or project on a Likert scale. To what degree are they willing to buy-in to your proposed project or recommendation(s)? With your presentation in mind, complete the below audience analysis.

Your turn

Use this information to structure your presentation.

1. Describe audience members. _____

_____

_____

2. Captive / Voluntary

3. Technical / Nontechnical

4. Desired outcome: _____

5. Degree of knowledge about _____

Little/None	Some	Unsure	Good Amount	Great Amount
1	2	3	4	5

6. Attitudes/predisposition concerning _____

Strongly Disagree	Agree	Neutral	Agree	Strongly Agree
1	2	3	4	5

As evidence of a thorough audience analysis, Jim did not simply repurpose the presentation he gave to his boss and the Mercedes-Benz Board when addressing the

Alabama Secretary of Commerce and Bibb County politicians. He instead opened with a narrative video of Mercedes-Benz International and their 20 years in Alabama. By first establishing a positive frame, he was able to easily introduce his purpose and persuasive appeal. Jim described the impact the new global logistics plant could have on Bibb County's economy, how a win for Mercedes-Benz would be a win for these politicians who were voted on by the governor and their constituents, and what Mercedes-Benz International needed to make that a reality. "Don't you want to help us help you?" He started with the end goal in mind, knew his audience, and brought only the most relevant information to the occasion.

Jim credits his success to not only creating strategic buy-in and connecting the organization's goal the audience's interests, but to seeking the feedback and advice from his boss and colleagues along the way.

> I did it, but it wouldn't have turned out that way without some feedback. My boss at the time had done this type of thing before. To is great credit he let me do it this time and gave me some excellent guidance along the way. If I hadn't solicited feedback, I believe it still would have been successful (after all, who doesn't try to create 600 jobs?), but not so easily and maybe not so much.

The foundation involves asking the right questions, but it is just the beginning of the overall process of building better presentations. In Chapter 5, we teach you how to construct the full presentation outline and place information into a format that is digestible for any audience.

4.5 CALLS TO ACTION

Before your next speaking opportunity, no matter how formal the situation, answer these key questions in Table 4.2:

TABLE 4.2   Key questions. Answer the below questions before constructing any message.

Key Questions	Your Answers
Who am I speaking to? Who needs the information?	
What is the purpose of my presentation?	
What is my desired outcome?	
What information is most important to my audience?	
Why should my audience care?	
When am I speaking?	
Where am I speaking?	
How will I present it? What types of types of visual aids would supplement my message?	

As you work through this chapter and your own next presentation, write your answers in the blank spaces.

As you practice being strategic with your information, you will become attuned to the degree to which others' have prepared. The next time you are in the audience, gauge how well the speaker answered the foundational questions prior to constructing the message. Would answering them have changed the way the message was delivered?

As we conclude Part 1 of this book, you have heightened your awareness of how communication can impact your profession. In the next four chapters, we put this awareness into practice and show you the steps to achieve excellent messages through organization, pitching ideas, creating visual aids, and cultivating charisma.

REFERENCES

1. Stovall, J.G. (2005). *Journalism: Who, What, When, Where, Why, And How | Pearson.* Boston: Pearson.
2. Stauffer, J., Frost, R., and Rybolt, W. (1983). The attention factor in recalling network television news. *J. Commun.* 33 (1): 29–37.
3. Wilson, K. and Korn, J. (2007). Attention during lectures: beyong ten minutes. *Teach. Psychol.* 34 (2): 85–89.
4. Scerbo, M.W., Warm, J.S., Dember, W.N., and Grasha, A.F. (1992). The role of time and cuing in a college lecture. *Contemp. Educ. Psychol.* 17 (4): 312–328.
5. McLeish, J. (1968). *The Lecture Method.* Cambridge, UK: Cambrdige Institute of Education.
6. Ed Enoch (2017). Vance plant to produce electric SUVs; will create 600 jobs. *Tuscaloosa News.*http://www.tuscaloosanews.com/news/20170921/vance-plant-to-produce-electric-suvs-will-create-600-jobs (accessed 13 November 2017).
7. Sunthonkanokpong, W. (2011). Future global visions of engineering education. *Proc. Eng.* 8: 160–164.
8. Chilcutt, A.S. and Chilcutt, A.S. (2009). Environment of a dental practice. *J. Am. Dent. Assoc.* 140: 1252–1258.
9. Nichols, R.G. and Stevens, L.A. (1957). Listening to people. *Harv. Bus. Rev.* 35 (5): 85–93.
10. Kramar, E.J.J. and Lewis, T.B. (1951). Comparison of visual and nonvisual listening. *J. Commun.* 1 (2): 16–20.
11. Gallo, C. (2014). *Talk Like TED.* New York: St. Martin's Griffin.
12. King, P.E. (2008). *The International Encyclopedia of Communication.* Wiley.
13. Gilbreath, J. and Eastman, K. (2011). The Credible Hulk: smashing student expectations through instructor crdibility. *Conference Proceedings Paper from LOEX 2016* (March 2011), pp. 85–86.

PART II

APPLY THE PROCESS

5

ORGANIZING AND OUTLINING YOUR PRESENTATION

The issue with most presentations is that people are trying to cram 10 pounds of information into a two-pound bag.
—John Baldoni, Author, International Speaker

* * *

Your presentation is like a road trip. As the speaker, you sit in the driver's seat and your audience are passengers along for the ride. They have no control over your route, whether you choose a circuitous or clear path. Your responsibility is to deliver them to their destination. If you can make the ride memorable, they will be more likely to accept your next invitation. But any uncertainty along the way creates barriers to effective listening. To make this a more engaging trip, you identify three points of interest along the way. People like to know where they are going. Clearly articulate the purpose of the trip, preview the route, and then guide them to the target destination. In this chapter you will learn how to

- understand the benefits of organization on audience reception and information retention,
- learn to build a structured outline in preparation for a presentation, and
- examine the formulas for crafting attention-getting introductions and strong conclusions.

Engineered to Speak: Helping You Create and Deliver Engaging Technical Presentations, First Edition.
Alexa S. Chilcutt and Adam J. Brooks.
© 2019 by The Institute of Electrical and Electronics Engineers, Inc. Published 2019 by John Wiley & Sons, Inc.

Creating a fully developed outline places the most valuable information in a format that unfolds logically. No matter the topic, the outline we share with you in this chapter has proven successful time and time again. The adage is to tell your audience what you are going to tell them, tell them, and then tell them what you just told them. This method involves a triptych, a device used to compose or present something in three parts or sections. While it may seem redundant, presenting information as three digestible parts works. Whether speaking for 10 minutes, 30 minutes, or 3 hours, organization is essential.

5.1 BENEFITS OF ORGANIZATION – DECREASING UNCERTAINTY

Organization improves comprehension, decreases uncertainty, and increases retention. As we launch into the organizational design of a presentation, you will see the power of three play out repeatedly. The mechanism of the triptych is used in the outline of three main points and the overall sequence of the introduction, body, and conclusion. This pattern primes the audience to receive and retain your message.

In the previous chapters, you were given the questions to ask yourself in advance of organizing the message. In a two- or three-minute pitch you do not have time for fully developed main points, but you must drill down to the importance of what you do, how the topic is relevant, and deliver a strong take-away. Chapter 6 details the process for developing powerful pitches. Longer presentations require a more developed outline with a logical arrangement of main points as seen below.

5.2 ENGINEERING THE OUTLINE

Consider the below outline for every presentation:

Introduction
a. Attention Getting Device
b. Link to Topic
c. Relevancy and Credibility
d. Preview Statement

Main Idea 1
a. Supporting Idea/Evidence
b. Supporting Idea/Evidence
c. (Transition to Main Idea 2)

Main Idea 2
a. Supporting Ideas
b. Supporting Ideas
c. (Transition to Main Idea 3)

Main Idea 3

a. Supporting Ideas

b. Supporting Ideas

c. (Transition to Conclusion)

Conclusion

a. Review Statement

b. Call to Action

On paper, this outline may look too simple for your complex project. When we coach individuals, we experience resistance and hear how unique and different their presentation is and that it could not possibly be distilled into this outline. They come to realize is that it can and are thrilled with the results. This formula works time and time again. The first step is to determine the specific outcomes you desire.

Your Turn

Write your topic in the line below and then consider your purpose statement. There is a reason the line is short. Make it simple and direct.

Topic: _____

General Purpose: (*circle one*): To inform. To persuade.

Purpose Statement: To inform the audience about _____.

Or

To persuade the audience to _____.

5.2.1 Informational Organizational Pattern

The body of the presentation accomplishes the *tell them* part of the adage we mentioned earlier. Break up your content into three main points that logically connect and flow cohesively. While many speech communication textbooks suggest having three to five main points, *the Rule of Three* seems to be most effective. The Rule of Three refers to the writing principle that threes create memory, humor, and satisfaction. It follows our natural desire for a beginning, middle, and end. Good things come in threes. Memorable slogans following this rule include, "Snap, Crackle, Pop," "Just do it," and "Life is Good."

Steve Jobs routinely applied the Rule of Three when unveiling products. When Jobs introduced the iPhone at MacWorld 2007, he began by telling the audience his purpose: to introduce Apple's revolutionary new products. He then previewed the aspects of the products he would be unveiling: "The first is a wide screen iPod with touch controls. The second is a revolutionary mobile phone. And the third is a break through Internet communications device." He broke the description of one product into three relatable parts. The three familiar concepts of the iPod, mobile phone, and

Internet guided how he described the concept of the iPhone. The iPhone included a myriad of features other than listening to music, making calls, and surfing the Internet, but had Jobs attempted to introduce and explain every feature, the audience would have been overwhelmed. By highlighting three major features, he captured their attention and ensured memory of the most revolutionary features.

Depending on your topic, there are informative and persuasive organizational patterns. Each provides a framework for a logical flow. The typical organizational patterns for informative presentations are chronological, spatial, or compare–contrast.

Chronological organization orders information by time or sequence. For example,

1. project's purpose and initial stage,
2. current state of project, and
3. next steps.

This organization is best for a narrative arc when following a timeline or in demonstrating how something works as you walk an audience step-by-step through a process.

A spatial arrangement of ideas is concerned with localized groups, size, or distance. Main points could be based on scale, as in large-to-small or small-to-large as well as geographic proximity as in near-to-far, east to west, or north to south. An example would be a topic that discusses a project's impact at the (i) company, (ii) national, and (iii) international levels. The chosen scale provides a framework for covering information as it relates to each area.

Finally, compare–contrast examines the similarities and differences of two subjects by highlighting three separate areas. For example, you might compare Design A to Design B in terms of (i) construction time, (ii) cost, and (iii) durability.

5.2.2 Put it into Practice

A professional who understands the power of a well-constructed message is Steve Butsch, the Senior Director of Sales Engineering at a dynamic technology company in Silicon Valley. With a master's degree in electrical engineering, Steve was recruited for a sales engineer position that required communicating complex technology to non-technical customers. In preparation, Steve observed how his colleagues conducted sales calls or presented at conferences. He asked himself, "what did this speaker do that made me *not* like the presentation? What did they do that *drew me in* and made it effective?."

As a successful sales engineer for the past 15 years, Steve defines an engaging speaker as someone who "moves through a presentation with ease–logically sequencing one concept to the next." His list of "what not to do" includes launching into technical details without introducing the topic or stating the purpose and having to back-track or continually pause to gather your thoughts. "I've seen presenters pridefully state 'I just finished these slides a few minutes ago.' I cringe when I hear this because that means I'll have to listen to someone who's trying to figure out their slides as they present them." One of Steve's favorite maxims is "Failing to prepare is preparation for failure." Recall our discussion in Chapter 2 that winging it does not work.

Your Turn

We asked Steve for examples of presentations he has delivered that would be organized differently according to the topic. For each scenario below, choose the best organizational method to arrange the main points of the presentation.

Scenario 1: You are representing a mobile tech company and presenting to clients about developing mobile applications for their products. The goal is informative. The main portion of your presentation will walk them through the sequence to follow when designing their first mobile app. Once a strategic and cross-functional team is assembled, they must work together to (i) identify the problem(s) the mobile app will solve, (ii) define its success metrics, and (iii) begin designing the app. Designing the app includes building it, measuring outcomes, and learn and revise.

> Circle one: Chronological → Spatial → Compare–Contrast

Scenario 2: You are launching a mobile platform and prepared a customer presentation outlining its benefits. You have information about how the mobile platform will positively directly impact stakeholders at various levels of the organization. Your stakeholders include corporate IT, the corporate app developer, and the corporate end user.

> Circle one: Chronological → Spatial → Compare–Contrast

Scenario 3: Your goal is to illustrate how your product is superior to the leading competitor's product. You must detail your product's design, help the audience visualize how the product operates, and clearly distinguish the product from its competition. You gather the information and create detailed tables that compare the products feature by feature. Key points for emphasis include battery life (e.g. "under normal loading, lab tests determined that a device running our software will last 32 hours whereas Competitor X only lasted 17 hours"), security, and user experience.

> Circle one: Chronological → Spatial → Compare–Contrast

Your Turn

If your next presentation will be informative, take the topic and purpose statement you have chosen and decide on an organizational pattern and three broad main points.

Choose the most appropriate pattern: Chronological, Spatial, Compare–Contrast

Main Point 1:

_____

Main Point 2:

_____

Main Point 3:

_____

5.2.3 Persuasive Organizational Patterns

When your goal is persuasive in nature, the objective is to adjust or change the group's attitudes. You must sway opinion, connect solutions to root causes, and end with an *ask* or recommendation that leads them to a specific action. The two primary organizational patterns for persuasive appeals are Problem–Cause–Solution or Cause–Effect–Solution.

In Problem–Cause–Solution, the first main point is the Problem. This is a good pattern to address problems people are less familiar with or for issues you must convince others to consider. You must fully define the problem to create shared meaning for a larger audience, what exactly *is* the problem, how is the status quo failing, and who is it failing. The second main point analyzes Cause(s) using evidence that attributes specific factors that created the problem. The third main point offers Solution(s) derived from what you identified as the main causes.

For topics that lend themselves to the Cause–Effect–Solution pattern, the first main point focuses on Cause(s), the second discusses Effects, and the third examines alternatives and arrives at the best possible Solution. Lead the audience to the solution once you have evidence to support, and verbally create a picture of how it provides relief for the problem. It is important to employ vivid language and allow your audience to visualize the solution's implementation.

A strong Close is key! Close with a specific *ask* that prompts the audience to act; which could include supporting your recommendation, providing funding or resources, or purchasing the product.

Your Turn

If your next presentation will be persuasive, take the topic and purpose statements you have chosen and decide on an organizational pattern.

Choose the most appropriate pattern: Problem–Cause–Solution Cause–Effect–Solution

Problem: _____

Cause(s): _____

Solution(s): _____

  Or

Cause: _____

Effect: _____

Solution: _____

5.2.4 Transitions

Transition statements connect each major section to the next. These statements are verbal cues that signal the progression of information. If we return to the road trip analogy, our passengers are reassured of the direction when they see (or hear) signs from one point of interest to the next. This is called signposting. Psychologically, it reduces ambiguity in the minds of the audience and increases their receptiveness to the message and memory of key points.

As an illustration, we will consider Steve's example of presenting to clients who want to create mobile applications. A transition statement between his first and second main point might be, "now that I have discussed how your team can identify the problem your mobile app will solve, the next step is to define metrics for measuring the success of the app."

An example from a persuasive speech would be a transitional statement that leads the audience from definition of the problem to the causes. As an example, "now that we have examined the problem, let's look at three root causes." These directional statements are critical for audience engagement and memory.

5.3 HOW TO BEGIN YOUR PRESENTATION

Presenters often fall into the routine of opening with, "Hello, my name is… Today I will be presenting on….," and close with, "That's all. Questions?" There is a better way.

In less than 30 seconds, initial impressions are formed and can influence another's opinion of your credibility and trustworthiness [1–3]. In these first few seconds, both your verbal and nonverbal communication contribute to the forming of these opinions. On the flip side, research indicates that people's attention spans can be limited to just eight seconds [4, 5]. The take-away is to make those first 30 seconds count and the best first impression possible.

Introductions and conclusions are golden opportunities to secure an audience's attention. Strong introductions take an audience by surprise and increase their receptivity to the message. Strong conclusions create closure by returning to an attention-getting device and delivering an all-important *ask*.

Be creative with introductions. Why begin like every other presenter, "Hello, my name is…"? An audience generally knows who you are and your topic because it is already listed in the agenda, program, or staring at them from the projected title slide. Do not waste those precious eight seconds telling them something they already know. Grab their attention. A strong introduction renders the audience receptive to the topic, establishes your credibility, and previews what you are going to tell them.

The formula for a strong introduction is Attention Getting Device + Link to Topic + Relevance (What's in it for me?) + Preview Statement. These four components are crucial to a well-developed introduction that secures the audience's attention, introduces the topic, creates relevance for the specific audience, and decreases uncertainty.

$$\text{Introduction} = \text{AGD} + \text{L} + \text{R(WIFM)} + \text{P}$$

An Attention Getting Device (AGD) is a language device that captures the audience's attention. Examples include shocking facts or statistics, narratives, quotes or rhetorical questions, even jokes if they are sure-fire winners. This is your opportunity to be creative and embrace enthusiasm about the topic by identifying something compelling. Think about the larger content of your presentation; if you are going to provide a lot of macro-level data, it might be compelling to begin with a narrative about a specific case. If you are talking about something micro-level, begin with something broad and narrow to your specific issue. The AGD becomes the first words you speak to open the presentation.

Imagine this: You step into the presentation space, whether behind a podium or in front of a conference table. Take a moment to let everyone's attention fall on you, and then open with your AGD. Our favorite AGD when beginning a public speaking workshop is "death, snakes, and public speaking." We are signaling that this is not going to be like every other workshop you have had to endure. No one is expecting "death, snakes, and public speaking" to be the first words they hear. We have just captured their attention.

Then, we link the AGD to our topic, "what do these have in common?" (Pause. People typically answer aloud, "Fear!") We continue, "they are listed as the top three phobias, however, for many Americans the fear of public speaking ranks above the fear of death. According to the National Institutes for Mental Health, approximately 74% of Americans experience anxiety associated with public speaking." This segment places the AGD in context to the topic.

Next, we answer the WIFM: "Research shows that across a variety of professions, those who are effective speakers experience a more rapid rate of career advancement and opportunity. You have all recognized the need to build these skills and are sitting among other professionals equally eager to learn and improve." In these first 30 seconds, we have grabbed their attention, introduced the topic, established credibility, and made the topic relevant.

Finally, the preview or the tell them what you are going to tell them: "Today's interactive workshop will equip you to become better speakers by learning how to craft effective messages, cultivate dynamic delivery, and calm public speaking anxiety." The workshop is organized into three sections as alluded to, and now the participants are less anxious about the order of information and how we will proceed. Overall, they are primed to pay attention and more agreeable to follow along.

Your Turn

As a recall activity, can you identify the elements of the introduction below? This example was taken from a presentation given by Annie Kary, a participant in an aerospace and mechanical engineering Research for Undergraduate (REU) program. Beside each section of her introduction, place the element it represents.

Introduction = AGD + Link to Topic + Relevance for + Preview Statement

_____Imagine trying to measure the thickness of a hair with a meter stick. How would you know the measurements have any accuracy or significance, if the scale that you are measuring on is so much bigger than the object you are measuring?

_____This summer, I will be working with Dr. Richard Barman at the University of Alabama Space Propulsion Observation and Testing Lab to shrink that meter stick down to a micrometer, but in the context of micro-Newton Hall Effect thrusters. Hall-Effect thrusters are electrostatic thrusters that use the Hall current generated in perpendicular electric and magnetic fields to ionize a gaseous propellant and produce thrust.

_____By 2020, it is expected that over half of all satellites launched will employ electrostatic thrusters for in-orbit maneuver. Hall thruster performance is a subject of great interest at aerospace research facilities ranging from collegiate CubeSat programs to national space agencies such as NASA and the European Space Agency for their favorable thrust abilities and lifetime, given their low mass. These thrusters can be easily used for delicate satellite positioning, drag compensation, and deorbit maneuvers that require adjustments too small for traditional solid- or liquid-propellant thrusters.

_____To discuss my research this summer, I will first provide you with some background on the thruster and different ways of measuring thrust on this scale. Then discuss what we hope to discover with this research, and how we hope to accomplish these goals. I will conclude by talking about the current state of the project.

5.4 HOW TO CLOSE YOUR PRESENTATION

Tell them what you just told them. The conclusion cements the purpose, recaps the main points, and finalizes the take-aways. This is the wrap-up and last opportunity to cue the audience toward a specific action or create buy-in. The point is to end with a bang not a whimper.

The formula for a conclusion consists of Thesis + Review + Relevance + Close/ Call to Action. Restate Thesis, Review the main points, touch back on Relevance to the specific audience, and deliver a strong Closing statement or a Call to Action.

$$T + R + R + Conclusions$$

A strong conclusion should parallel with your AGD. For example, if you began with a compelling fact or statistic, end with another compelling fact or statistic. If you opened with an inspirational quotation, end with another quotation that reflects your call to action. In a persuasive presentation, a conclusion must include a clear statement of what action you would like the audience to take and how they can engage in taking that next step. Remember Jim Hans from Mercedes-Benz International in Chapter 4? He described his closing method for presenting possible solutions, driving his audience toward a specific recommendation. Jim created the final slide with all recommendations listed, placing a check next to the one he would like the audience to select and verbally made the final pitch for that specific action.

Our workshop conclusion mirrors our introduction: "Today, we have helped you become more effective speakers by teaching you how to 1) craft effective messages for a variety of audiences, 2) cultivate dynamic delivery, and 3) calm public speaking anxiety. Along the way, you have engaged in activities where you delivered both a shorter pitch and the introduction to a full presentation and have received feedback from our facilitators that helped you improve. Out of the tops phobias, death may be inevitable, snakes largely avoidable, but public speaking is to be embraced as an opportunity to share your ideas and advance your career!."

5.5 PREPARATION OUTLINE

The process presented in this chapter focused on developing a full preparation outline. This is a full-sentence outline that follows the Introduction, Body, and Conclusion format with development of the three main points with subpoints and transitions that segue from one major section to the next. The full preparation outline format is found in the Supplemental Chapter at the end of the text.

Once your outline is complete, the next step is to understand the role nonverbal communication plays in the delivery of the presentation. This includes personal nonverbal communication, such as gestures, movement, vocal variety, and facial expressions as covered in Chapter 8 as well as your visual aid design and implementation as covered in Chapter 7. In the end, practice will be the key to improving delivery and reduced anxiety. Visual aids are effective when used appropriately and with the material and audience in mind. The goal is to complement your spoken message and increase audience engagement and information retention.

5.6 CALLS TO ACTION

The road map has been laid out before you. This outline process will be useful for any presentation, and we have placed the full process in the Supplemental Chapter for your repeated use. Now that you have identified the purpose of your presentation, be creative and craft each of the introduction and conclusion components.

Create an Introduction

Topic:

_____

Attention Getting Device:

_____

_____

Link to Topic:

_____

_____

Relevance (WIFM):

_____

_____

Preview of Main Points

_____

_____

Create a Conclusion

Thesis:

_____

Review of Main Points:

_____

_____

Relevance:_____
Close (bring back to AGD) or Call to Action:

_____

_____

REFERENCES

1. Begrich, L., Fauth, B., Kunter, M., and Klieme, E. (2017). Wie informativ ist der erste Eindruck? Das Thin-Slices-Verfahren zur videobasierten Erfassung des Unterrichts. _Z. Erzieh._ 20: 23–47.

2. Gilbreath, J. and Eastman, K. (2011). The Credible Hulk: smashing student expectations through instructor credibility. *Conference Proceedings Paper from LOEX 2016* (March 2011), pp. 85–86.
3. Ambady, N. and Rosenthal, R. (1993). Half a minute: predicting teacher evaluations from thin slices of nonverbal behavior and physical attractiveness. *J. Pers. Soc. Psychol.* 64 (3): 431–441.
4. Bradbury, N.A. (2016). Attention span during lectures: 8 seconds, 10 minutes, or more? *Adv. Physiol. Educ.* 40 (4): 509–513.
5. Gee, P., Stephenson, D., and Wright, D. (2003). Temporal discrimination learning of operant feeding in goldfish (Carassius auratus). *J. Exp. Anal. Behav.* 62 (1): 1–13.

6

PERFECTING YOUR PITCH

Getting better at sharing ideas means making the complex simple and the simple interesting. In Chapter 5, you learned how to outline a full presentation, but a shorter pitch is useful when seeking funding, updating stakeholders on your progress, or networking with potential clients. In this chapter, you will learn how to

- identify the parts of a successful pitch,
- relate the elements of a compelling narrative to your experience, and
- construct your own framework for your next project.

Let us begin with an example of a pitch that fell flat. During a product introduction meeting at a major construction equipment manufacturer, Adam sat in a room full of the smartest people in a Fortune 50 company. They were tasked with achieving design innovation to meet new government emissions requirements on a multibillion-dollar product line. When team leads updated the company's marketing and public affairs leaders, they launched into intricate detail about every decision they made throughout the process. They discussed every test, including the ones that failed, showing graph after graph before getting to the final recommendations. After the presentation, no one could properly summarize the updates.

Engineered to Speak: Helping You Create and Deliver Engaging Technical Presentations, First Edition.
Alexa S. Chilcutt and Adam J. Brooks.
© 2019 by The Institute of Electrical and Electronics Engineers, Inc. Published 2019 by John Wiley & Sons, Inc.

While technical professionals are process-oriented and enjoy examining intricate details, many audiences need the message to be broken into actionable bits that help them understand where they play a part. People want to know what they are supposed to *do* with the information. From Adam's perspective, the engineers neglected to synthesize the information and lead with what mattered most to the marketing and public affairs leaders.

A pitch drives your audience to a specific action through a brief snapshot of a larger idea. This information is delivered in a compressed period, typically one to three minutes. For a pitch to be successful, it is best to think of the ideas you want to present as an argument. You must first understand how to build a compelling argument, and then assemble pieces that bring your argument to life. We provide you with a template for building a narrative that illustrates a tangible example and offers the framework for the next time you pitch a project. Making the case for your project involves three elements: an essential truth, tangible examples, and a call to action.

6.1 LEAD WITH MEANING

First impressions are powerful and long lasting, yet we often lead with the things that do not necessarily matter. When people outside of your professional sphere introduce themselves using their job titles, do you find yourself wondering, "what does that mean"? In Chapter 4 we discussed the value of understanding your audience. Too often, we make our communication about *us* instead of furthering the needs of our audience.

You need to lead with meaning. Leading with meaning is about putting the most essential piece of information – the nugget and overall synthesis of the entire presentation – into a clear and recognizable statement. It is distilling the complicated web of ideas, facts, and figures into a simple sentence your audience recognizes as the main point. From there you can expand your ideas, adding more details and helping bring your audience into shared understanding.

6.2 START WITH AN ESSENTIAL TRUTH

Lead with an essential statement that encapsulates whatever you are proving. While your audience matters, getting to a "Yes!" means taking it a step further and considering the audience of your audience. Research into what makes a message go viral suggests the success of a message depends on how easily your idea can be shared by influential individuals [1]. It is not the audience of the project proposal that matters, but how easily that audience can share your idea with *their* audience. This means creating something that is concise enough to be memorable and repeatable. Getting to an essential truth requires drilling to the core of your strategic objective. What action do you want the audience to take? How can you articulate the value you bring?

Your Turn

Think of a project you are currently working on. What is the essential driver of that project? Drill down to the key competitive advantage or value proposition of the project.

Take a minute to write this down.

_____

_____

_____

When we pitch ourselves as communication consultants, we could say, "we help individuals gain a competitive edge by increasing their public speaking and presentation skills." That is a mouthful. Moreover, there is nothing particularly active about that idea. If you put down this book and we asked you what we did, you would probably come back with, "something about speaking." Getting to the essential truth is about creating an active idea you can express in a few memorable words. Instead, we open by saying, "we help companies develop their talent to make their ideas more effective." Our essential truth is that we enjoy helping others succeed.

6.3 USE EVIDENCE TO TELL YOUR STORY

It is worth considering arguments as something broader: influence. When it comes to influence, humans are no longer motivated by only facts and figures. In 1984, communication scholar Walter Fisher argued that while previous eras of humans relied solely on reason and rational argument, the modern era requires more types of evidence. Twenty-first century communicators must frame their argument as a simple yet powerful narrative.

Stories are powerful because they are easy to remember. Cognitive psychologists who study fluency note that the ideas that require the least amount of cognitive effort tend to be the ideas people transmit more often [1]. Narratives are the reason humans often act against their own self-interests, and one of the reasons why advertisements campaigns are effective. For example, we have known for decades that smoking is detrimental to our health. Yet, figures like the Marlboro Man evoked a cowboy narrative of rugged masculinity and a connection with the outdoors that resonated with the American archetypes [2].

Narratives are just as powerful in the workplace, and companies that can tell their story to consumers are more successful. In endeavors of entrepreneurship, investors are swayed by the effectiveness of a company's narrative. Amazon's Jeff Bezos made headlines for announcing his company had banned linear PowerPoint presentations in meetings [3]. Instead, employees were required to create shorter narratives that told a complete story [4]. Narratives are the ways in which societies articulate deep seeded moral beliefs and values [5, 6]. Therefore, to effectively increase the influence of your ideas, you must put them into coherent and compelling narrative examples.

6.3.1 Building Compelling Narratives

All stories share common elements. Conventionally, we come to expect stories to have a beginning, middle, and end. Any story you read likely has a protagonist attempting to accomplish a goal, and who encounters obstacles before that goal is resolved. While this gives us a basic framework, it leaves a lot to be desired in terms of creating something functional for a business meeting.

Let us reframe the idea of a beginning, middle, and an end to something more suitable for personal, project, and product narratives – from something, through something, to something. Hiring managers often train interviewees to answer behavioral questions in a similar form using what is often referred to as the STAR method, which stands for

- Situation
- Task
- Action
- Result

Here you can see the "from something, through something, to something" idea played out in a formula anyone could use when building a compelling narrative. It allows you to convey the most essential information your audience needs to understand, the framework of the solutions you have developed, the new product specifications, or design changes.

A narrative must also put a face to the work. In the case of most engineering communication situations, you will want to form narrative examples around a few strategic protagonists, such as the customer and the stakeholder. Instead of thinking through how the recent design change impacted your team, focus on how it impacts the end users and their situations, tasks, actions, and results. Can you frame a story of how customers engage with a product, and how that engagement could be improved by implementing your solution? In the case of stakeholders, consider their expectations and provide examples that resonate with their experiences.

6.3.2 Applying the STAR Method

Building compelling narratives with the STAR method takes you from something, through something, to something. Convey the most essential information your audience needs to understand your experiences. Once you start framing anecdotes in this way, you will realize how unfocused and disorganized most people are when they attempt to tell a story. Each story constructed with the STAR method should take 30–60 seconds total. These scenarios are used as your tangible examples in the upcoming pitch formula.

Your Turn

Use the space below to develop your first narrative:

What was the initial **situation**?

Situation:

_____

_____

_____

_____

What **task**(s) were you assigned?

Task(s): _____

_____

_____

_____

What **actions** did you take to complete the task(s) successfully?

Action: _____

_____

_____

_____

_____

What were the **results**? How was the situation resolved?

Result: _____

_____

_____

_____

_____

Now that you understand the elements of a good story, build a bank of them to mobilize for typical scenarios.

Adam used to work with competitive public speakers at the national level who entered competitions in impromptu speaking. These speakers had 30 seconds to develop a commanding and compelling seven-minute speech based on a selected prompt. As they move through national competitions, speakers cannot use the same material over and over and must constantly use new ideas and examples. In this environment, it is common for the elite speakers to develop story banks that they can mobilize for the variety of topics they must address.

Similarly, we encourage professionals to keep a small notebook of examples in their office to help remind them of key areas they are already comfortable talking through. As you experience key breakthroughs in the project, or finish particularly memorable achievements in your work efforts, be sure to record a simple narrative reminder for yourself. Building tangible examples of stories increases your speaking effectiveness and allows you to come across as more conversational.

6.4 END WITH THE CALL TO ACTION

Most pitches build to a call to action. As mentioned earlier, you want to establish a specific purpose behind your presentation. An effective call to action offers a clear direction to your audience of exactly what you want them to do. This also shifts the attention from you and puts it back on the people you are targeting. In the case of your career elevator pitch, you might end with, "so tell me more about the needs you have in the research and development business unit." The goal to audiences is a concrete directive that they can act on. When we put it all together the framework looks like this (Figure 6.1):

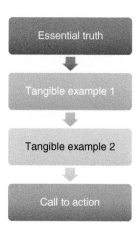

Figure 6.1.   Elevator pitch framework include four blocks that build to a specific call to action. The boxes flow from top to bottom beginning with essential truth, then tangible example 1, then tangible example 2, and ending with call to action.

We have crafted this formula for an effective pitch, but, like all formulas, it is best when adapted to fit within given circumstances. Use this as a framework to build the specific arguments and ideas you wish to communicate. Over time, you will have developed a cogent set of talking points that you can employ for any speaking occasion. Thus, you will be able to make the most of every communication opportunity.

6.5 CALLS TO ACTION

Use the space below to map your own pitch. Use this model as a keyword outline, but make sure you practice explaining each example as you go.

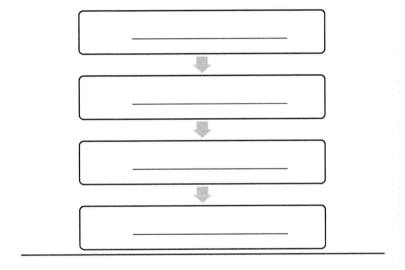

REFERENCES

1. Thompson, D. *Hit Makers: The Science of Popularity in An Age of Distraction*. New York: Penguine Press.
2. Starr, M.E. (1984). The Marlboro man: cigarette smoking and masculinity in America*. *J. Pop. Cult.* 17 (4): 45–57.
3. Cao, S. (2018). Why Jeff Bezos doesn't allow PowerPoint at Amazon meetings | observer. *New York Observer.* https://observer.com/2018/04/why-jeff-bezos-doesnt-allow-powerpoint-at-amazon-meetings/ (accessed 10 May 2019).
4. Bezos, J. (2017). 2017 Letter to shareholders. *The Amazon Blog Day One.* https://blog.aboutamazon.com/company-news/2017-letter-to-shareholders/ (accessed 10 May 2019).
5. Rowland, R.C. (1987). Narrative: mode of discourse or paradigm? *Commun. Monogr.* 54 (3): 264.
6. Fisher, W.R. (1985). The narrative paradigm: an elaboration. *Commun. Monogr.* 52 (4): 347–367.

7

VISUALIZING YOUR MESSAGE

When the eyes and ears compete, the eyes win. Likewise, when a presenter shows an audience slides packed with text-heavy bullet points or complex graphs, charts, and equations, the audience is reading and processing rather than listening. You have invested time and energy preparing the message, now it is time to ensure return on investment by creating valuable visual aids. In this chapter, you will learn how to

- recognize the various types and use of visual aids;
- employ tips and tricks for interactive white board and flip chart use; and
- apply effective slide and handout design.

A productive place to begin is to identify what you *don't* want to do. Think about the most recent presentation you have sat through where a presenter's use of visual aids was ineffective. Take a moment to list your most common presentation annoyances.

_____

_____

_____

Engineered to Speak: Helping You Create and Deliver Engaging Technical Presentations, First Edition.
Alexa S. Chilcutt and Adam J. Brooks.
© 2019 by The Institute of Electrical and Electronics Engineers, Inc. Published 2019 by John Wiley & Sons, Inc.

Information about slides as visual aids:

- The purpose of the slide design template is to allow user to bring color, visual text, and images of objects or data into the presentation.
- Templates are pre-formatted, easy to manipulate, and allow for embedding of video, audio, or excel files.
- While slides are valuable as a general tool, they quickly become a crutch for speakers.
- An estimated 30–35 million presentations are delivered daily using PowerPoint with an average of 40 words per slide.
- *PowerPoint Karaoke* describes presenters who cannot begin without the first slide in place or who look as though they are seeing the next slide for the first time.
- Instead, adopt a strategic and creative approach.

Figure 7.1. Example of common PowerPoint layout. Presenters misuse the standard PowerPoint template by creating text-heavy slides. Visual aids need to include minimal text and a sharp image or graphic.

Typical annoyances include presenters reading directly from text-heavy slides, distracting fonts, background colors, motion, or too much data.

If we had the slide in Figure 7.1 displayed behind us as we spoke, would you be reading the slide or listening to us? Think outside of the box(es) provided by slide templates and boycott the overuse of bullet points. There is no evidence to support the 7×7 or 6×6 rules, as in number of lines or words per line. The purpose of visual aids is to enrich spoken information with images that illustrate a concept. Visual aids engage an audience by using multisensory channels that interact with both sides of the brain. A talk-and-show approach to presenting information clarifies meaning and reinforces memory.

We encourage you to adopt the mindset that creativity thrives within constraint. A good example of operating within constraint includes graduate students who compete in the Three Minute Thesis (3MT).

The 3MT is an international university-level competition that requires participants to deliver a research presentation to a general, non-technical audience in three-minutes or less using just one static slide. The competition originated from the concern that those with the world's most advanced knowledge were incapable of articulating those ideas to general audiences. The 3MT builds on the premise that if you met someone with the ability to invest in your research, would you be able to explain it effectively? If you have ever written a thesis or dissertation, or spent enormous time on a project, imagine accepting this challenge. How would you create a compelling three-minute overview of your project? What would you place on that one slide? Would you use the slide to relate to the audience, or to display important and visually interesting aspects of your research?

Christopher Ceroici, a PhD candidate in biomedical engineering at the University of Alberta, struggled with visual aids until he learned how to make those visuals talk with him rather than for him. His research focuses on developing 3D ultrasound imaging systems comprised of micromachined transducers called CMUTs. This method creates more portable and sensitive medical diagnostic tools and assists with the detection of cancer and other diseases. As a professional challenge, Christopher entered his university's 3MT competition.

Successful competitors approach the challenge from many different angles. For example, rather than using a scanning-electron microscopy image of several CMUTs, Christopher demystified a complex topic with a narrative. He displayed an image of Dr. McCoy and a patient aboard the Star Trek Enterprise. In the depiction, Dr. McCoy uses a small device to scan the patient sitting on an examination table. In his presentation, Christopher began – "most people have heard of the tri-corder. The seemingly magical device that the crew of the Enterprise use to scan for medical abnormalities with a simple wave. But for patients who have undergone medical imaging in the real world, they know this is a far cry from the drugs, dye injecting, fasting, and health risks associated with X-Ray or MRI imaging."

Christopher then segued into his research on ultrasound imaging technology comprised of micromachined transducers. This technology could revolutionize many current medical imaging constraints. His presentation approach was unique and incredibly effective.

Recall Chapter 2 when we discussed the limitations of time as well as the idea that creativity thrives within constraint [1]. Visual aids can be impactful if designed creatively and strategically. Simplicity is the go-to rule; you want visuals to reinforce, not compete with, your spoken message. We asked Christopher what he learned from the experience:

> The competition gave me an opportunity to dig through the technical aspects of my research and extract the information that really matters and would have an impact on someone unfamiliar with the field. Forcing myself to discard unnecessary details and complexities was extremely challenging at first. Eventually, I realized that refining a complex topic down to a simple story or metaphor gives the audience something to hold on to throughout the presentation. I now try to use these techniques in everything from technical presentations at conferences to explaining my research to my family.

Most speakers have more than three minutes, and not all presentations require slides. In this chapter, we examine the benefits of visual aids as well as the varieties, uses, and methods for effective construction. We consider how and when to use slides, handouts, white boards, and flip charts.

7.1 UNDERSTANDING WHY VISUALS WORK

Twenty-first century professionals operate in a visual society. People are constantly bombarded with stimulation. Rather than adding to this unnecessarily, presenters need visuals that make abstract concepts more concrete. Remember that oral communication (unlike written) relies on sensory input. The more you effectively engage a variety of senses, the greater the learning outcomes.

The power of images, visual representations, and demonstrations in teaching and learning are indisputable. Edgar Dale's *Cone of Experience*, published in the 1940s, continues to be the basic guideline for learning effectiveness, stating that

People remember about 10 percent of what they read, 20 percent of what they hear, 30 percent of what they see, 50 percent of what they see and hear, 70 percent of what they say, and 90 percent of what they say as they do a thing. [2, p. 432]

Your verbal message creates awareness. Visualization of your information creates context. Control is given to the audience when you allow processing to occur. Why settle for 20% when you could ensure 50% audience retention by adding well-designed visuals? To increase retention beyond 50%, use multisensory channels. Research suggests that visual aids increase audience interest, overall material recall, and perceived speaker credibility [3–5]. Through visualization of facts, figures, graphic data, images, video, tangible objects, or demonstrations, the audience is given evidence that confirms the spoken message.

Beyond the well-constructed visual aids within the presentation, ending a presentation with a visual summary, as seen in Figure 7.4, also increases comprehension, recall, and correct actioned responses [5]. A visual summary may include an illustration of the sequence or linear flow of the information covered, or a diagrammatic process with short captions. This concept brings us back to the *tell them what you told them* portion of your message from Chapter 5. Here, however, you are *showing* them what you have told them. Table 7.1 lists a variety of visual aid types and their intended uses.

TABLE 7.1   Types of visual aids and general uses.

Visual Aid	Uses
Slides	Projection of slides containing images, text, color, and embedded links, usually in a sequence following the material.
Overhead projector	Demonstrate work solving problems, illustrating, etc. in real-time using transparencies. May have pre-prepared transparencies with partial work.
White board	Writing out points, showing work solving problems, illustrating in real-time. Marker size depends on size of room.
Flip chart	Making lists, brainstorming, showing work in small groups.
AV – Video/Audio	Show recording of movement or relevant topic video or listen to audio-only of specific sound.
Physical object	Exhibit physical item for audience to see, touch, feel, and interact with subject matter.
Demonstration	Show or display process through physical motions or manipulation of an object.

Listed in the left column of this table are the most common types of visual aids. Beside each type is a description of the ways presented.

7.2 DESIGNING YOUR SLIDES

Professionals use visual aid software as a base design. You have many design options, including Google Slides, Keynote, Prezi, and PowerPoint. The company Forethought developed PowerPoint software for Macintosh in 1987 (and originally called it Presenter) [6]. It has been widely available as part of the Microsoft Office Suite, and is the go-to for slide design. The slide design templates allow users to bring color, visual text, and images of objects or data into their presentations and reinforce the spoken word. The templates are pre-formatted, easy to manipulate, and allow for the embedding of video, audio, or spreadsheets.

While slides are a valuable tool, they can become a crutch for speakers who use them as visual outlines that they read to audiences. An estimated 30–35 million presentations are delivered daily using PowerPoint with an average of 40 words per slide [7–9]. To illustrate another successful approach to the 3MT challenge, look at Sushruta Surappa's slide design in Figure 7.2. Sushruta is a graduate research assistant at the Micro-Machined Sensors and Transducers Lab in the Woodruff School of Mechanical Engineering at the Georgia Institute of Technology. His slide contained two illustrations with labeling and two images.

The slide is not crowded. It uses white space and shows the *what* and *how*. In explaining his design approach, Sushruta said

> The idea was to convey the essence of the talk without unnecessarily overcomplicating it. I wanted to have illustrations that captured the three main ideas of my talk – the need for the technology, the implementation and the final product. To keep it simple, I wanted

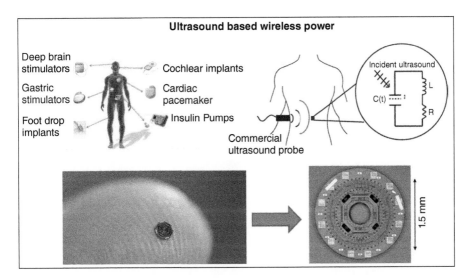

Figure 7.2. 3MT competition slide titled: Ultrasound based wireless power. Source: Image of presentation slide used to compete in Georgia Institute of Technology's 2017 3MT Competition. Used by permission of Sushruta Surappa.

to limit the number of images to three or four at most. I also didn't see the need for text as I didn't want the judges reading the slide and getting distracted as I gave the talk.

To create effective slides, we will discuss how to storyboard your ideas before providing you with finite guidelines.

7.2.1 Storyboard Design

When given unlimited freedom, how many of you spend way too much time glued to the computer screen working on a slide deck? David Daughton, a physicist and applications scientist, told us the story of one presenter who spent an entire weekend designing an intricate PowerPoint diagram of a complicated deposition. While the technical audience may have ultimately decoded and understood it, the complexity of the slide diverted attention from the speaker and was an enormous time suck for the speaker prior to the presentation.

To combat this pitfall, David approaches slide creation by first storyboarding his talk on paper or a whiteboard. This technique can also be done with index cards. For example, in preparation for a 10-minute talk, David draws 10 boxes and fills them with an image or diagram for each key concept. "This way, before I start making slides, I see the story I am trying to tell. It also helps me avoid making too many slides I don't plan to use." By identifying the larger narrative, non-essential information is more easily recognized and eliminated. Storyboard in hand, David constructs his slides. "I tend to focus on making clear diagrams. If available, use a marketing department or graphics team to make the diagrams better. Good diagrams make your speaking job easier." You cannot draw something you do not understand. Images are honest.

Your Turn

Think of an upcoming presentation and write or draw the first two key concepts. Create a simple statement or illustration in each box.

For a 10-minute presentation, you may only need five or six slides. It all depends on the time you expect to spend on any given concept. Therefore, the number of slides depends on what is necessary to show, emphasize, or explain. Once you have storyboarded the presentation, open your preferred computer software and transfer your ideas. As David suggested, good quality images make a difference.

When images are not enough, consider using the Assertion-Evidence slide style created by engineer Michael P. Alley [10]. This format creates a more focused presentation and allows for a conversational style of speaking. The Assertion-Evidence method's design has been shown to have positive effects on learning and retention [11]. Engineering students taught via Assertion-Evidence slides reported a decreased perceived cognitive load (i.e. they did not feel overwhelmed with information, combatting cognitive backlog), possessed stronger recall skills, and increased comprehension of complex concepts.

To follow Assertion-Evidence design, (i) build each slide based on one key message as an assertion (full sentence), (ii) support that message visually, and (iii) explain the evidence conversationally [12]. Phrase your assertion as a complete sentence at the top of the slide inside the header box. Below the assertion, place the visual evidence to support the claim. This may include photographs, simplified data points data, or diagrammatic processes. You may use more than one image if necessary.

7.2.2 Keep it Simple

It is easy to get caught up in your own work. A complex image that makes sense to a team focused on a project does not always translate to an audience seeing the information briefly and for the first time. Additional considerations for slide design have been gathered and listed in Table 7.2.

Let us examine how one group worked through this process of simplification. Alex, a mechanical engineering major and the mechanical lead for his EcoCar team,

TABLE 7.2   Dos and Don'ts of slide construction.

Do Use	Do Not Use
Claim or assertion at top in sentence format	Text-heavy slides including overuse of bullets, paragraph format
Clear, high quality images and graphics representing key concepts	Busy diagrams or graphs with too many labels or competing data
A key equation that summarizes point	Multiple equations on single slide
San serif font	Multiple types or stylized fonts
Dark background with white text	Neon backgrounds or pastel color fonts
Larger fonts (≥24)	Small font (≤20)
Consistent background and colors throughout	Text or slide animations for emphasis
Attributions of images if not original	Too many images on a single slide
Embed videos/include audio if sound is applicable	Videos linked to outside source

The left column of this table contains a list of constructive considerations, the Dos, when crafting your slides. The right contains a list of negative presenter habits that create a busy or confusing slide, the Don'ts.

developed the summary slide for his project. In describing his team's original slide (see Figure 7.3), Alex said, "this is the code we included in our first draft of our presentation. Basically, it's a mess of multiplying and adding things."

Imagine this slide projected for a limited amount of time as the speaker presents. Would this image make sense to someone sitting in the tenth row? Are they listening to the speaker or trying to read the labels and follow the diagram's flow? Thankfully, this was the initial draft.

The team's final design was much improved, but we included the first draft to stress that sometimes your first idea is not your best idea. In describing the final slide design (see Figure 7.4), Alex said, "This is what we ended up showing at competition. We sat down and drew up what each step of our code did and completely replaced the block diagram mess with a flowchart. The little picture is a basic description of what we did at each wheel. The outcome is less technical, but anyone with an engineering background would understand our approach."

The visualization of your information creates context. Control is given to the audience when you allow processing to occur. Talk about your concept by introducing and discussing the claim. Then, visualize the concept for your audience and build in a pause to let the graphic sink in. Say, show, sink in.

For discussion that does not necessitate a slide, place a blank, black, or dark background slide between the visual slides as a placeholder. This keeps the audience from focusing on a previous concept and shifts the focus back to the speaker.

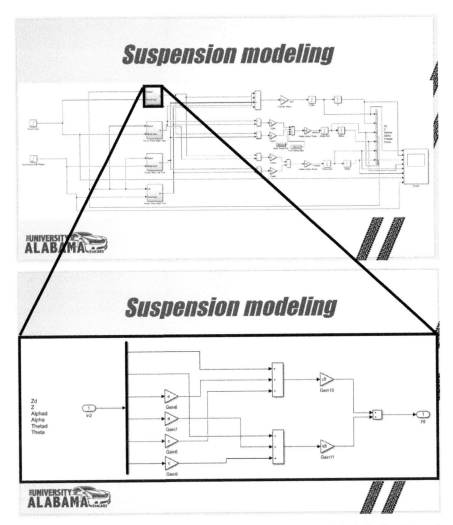

Figure 7.3.   Image shows the first draft of workflow design to include in The University of Alabama EcoCar3 Mechanical Team's final presentation. Source: Image used by permission of Megan L. Hathcock, Alexander J. Matlock, Joshua M. Penn, and Miranda M. Tanouye.

7.3 HANDLING HANDOUTS

Handouts are one of the most underused visual aids for retention of information. The normal bandwidth of an audience to listen, interpret, and retain information from a spoken message is compounded by where you are in a line-up of presentations. A handout provides the audience with detailed information for improved comprehension and the ability to review it as needed.

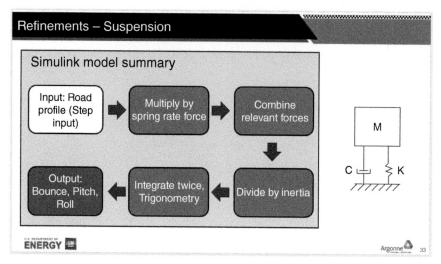

Figure 7.4. EcoCar competition final slide design of flowchart. Slide illustrates graphic using condensed workflow design. Source: Image used by permission of The University of Alabama's EcoCar 3 Mechanical Team, including Megan L. Hathcock, Alexander J. Matlock, Joshua M. Penn, and Miranda M. Tanouye.

Please note that handouts are *not* a printed copy of your presentation slides. When you do this, you signal to the audience that you could have e-mailed the slides rather than have them sit through your presentation. Handouts should overview the information presented as well as include details related to the findings or proposed actions.

During the presentation, discuss the slide information first and then refer to the specific table, figure, or text at the appropriate time – e.g. "As you see on this slide, the most surprising finding was … in your handout, see Figure 1 for more detailed data." Verbally focus the audience's attention to the most important finding on your slide while providing them with proof of the larger scope of work should they desire it. A one-sheet handout, even if printed on both sides, is sufficient. Keep the sequence of the information in line with the oral presentation but make the material supplement your spoken delivery. Ensure that your graphs are large enough to read easily with the labels and visual codes included. Use nothing smaller than an 11-point font. At the beginning of the presentation, state that you will refer to items in the handout at the appropriate time.

We create handouts for our own talks and training sessions. These handouts provide context visually and encourage interactivity. We do this by constructing them according to the presentation's main points, reinforcing key data, including spaces for people to write or answer questions, and providing a list of take-aways and contact information. For those of you who present at conferences, the people who have opted to attend your session are by nature interested in your topic. A handout makes the presentation memorable and acts as a calling card.

Handouts have saved us more than once. If technology fails (and, at some point, it inevitably will), a handout will keep you focused on the flow of information and provide visualization and context for your audience. If you are in a line-up of speakers who arrive to find that the AV does not work, you will be the one who can proceed with little to no trouble. The presentation may be a little off-script but you still have a general outline and your audience has visual confirmation.

7.4 PHYSICAL OBJECTS/DEMONSTRATIONS

Show and tell works for adults too. When you have a physical object that is transportable, you allow the audience to view its texture, size, or composition. You are engaging their senses in a real and tangible way. Depending on the object, the audience may have the opportunity to interact with it. If the object is small and sturdy, it could be passed around. Be careful of the timing with passing an object around. Once people begin to engage and examine it, the focus is off you and what you are saying. Describe the object prior to passing it or revealing it. Keep with the rhythm of say, show, sink in.

Presentations are more dynamic if you demonstrate a process or show how something operates, allowing your audience greater insight and to ask more in-depth, process-oriented questions. Planning for a demonstration means thinking through clear language that describes an easy-to-follow process. Language choices need to be vivid and precise. Write the steps in sequential order and then visualize physically demonstrating those steps. Include language that signals transitions between actions and purposefully include transition statements like, "Now that we have … the next step is to…Finally, …"

Practice is necessary when including a demonstration. Take into consideration the space you are given to present, the size of the objects, the placement of the objects on a table or central location, and the ability to demonstrate so people can view what you are doing. Demonstrations may also include audio visual support using images or short video clips. We will address practice in more depth in Chapter 9. We strongly suggest performing a trial run for a test audience member and ask for feedback to ensure process-oriented clarity and receive feedback. Ask the following questions after your trial demonstration:

- How easily could you follow the progression of the steps in the process?
- Were there any portions that needed clarification?
- Was my language clear and descriptive?
- Were you able to view the demonstration easily?
- What suggestions do you have?

7.5 REAL-TIME WRITING AND DRAWING

There are occasions, especially in team settings, where brainstorming and collaboration are essential, where a speaker or facilitator needs to show work or engage with a
group to ensure understanding or gather feedback in real time. Writing text, working
problems, or drawing diagrams and figures bring an interactive element to a presentation as the audience is paying attention to the work in progress on a static medium
like a white board or flip chart.

The benefits of these static display methods engage you in active processes;
provide natural pauses in the presentation as you write or draw, allowing audience to
process along the way; offer dependable low-tech visual aids; and encourages a more
interactive presentation, leading to questions and greater feedback. There are many
options in our tech-savvy work environments, but the staples we cover below are a
sure thing. For this section, which required artistic expertise, we consulted visual
thinker and agile coach Yuri Malishenko.

7.5.1 The White Board

Whether in a classroom or boardroom, the white board is a go-to platform for writing
and drawing ideas as well as engaging with the audience. The size of the room and
the group is an important consideration when creating visuals to be seen effectively
from any seat. If you are in a larger space where people may be sitting at the back of
the room, go online and purchase the dry-erase markers that have a 15 mm tip rather
than the standard size found in grocery stores.

Drawing may not be your strength. That is OK. For grids, sticky notes are a perfect 3×3 size to trace for each square within a larger grid (see Figures 7.5 and 7.6).
You may need a larger template for bigger audiences. Print words rather than using
cursive and write some words on the board before the presentation to ensure they are
visible from the back of the room. Gauge the size of the text using the width of one,
two, or three fingers. This gives you a quick test during the presentation to keep your
writing a consistent size.

Drawing a dashed or broken line indicating direction or flow? Draw a straight
line first, then use your finger to remove sections of the outline and it produces a
nice-looking dashed line. It is much more difficult to draw a straight dashed line to
start with.

Remember, a white board is forgiving. Make a mistake? Erase it. Think of a new
idea? Write it out. Double-check that you have access to an eraser and good dry-erase
markers in the most appropriate colors. Use darker markers for primary text and
shapes and colored ones for emphasis or shading. Stay away from pastel or lighter
colors. Too many colors are distracting, so stick to one or two for emphasis. 3D
effects applied using sticky notes prepared before-hand with numbers of drawings or
colored circular magnets add a pop of color and draw the eye to an important point
(see Figure 7.7) [13].

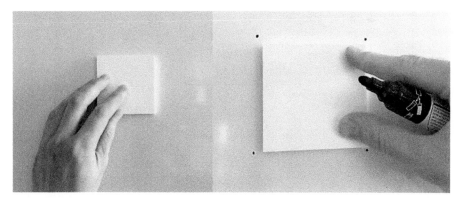

Figure 7.5.   Image of presenter using a small square note pad as a template to mark equal corners for the construction a grid design on a white board. Source: Used by permission of Yuri Malishenko [13].

Figure 7.6.   Whiteboard grid outline. Presenter creating larger grid on white board. Source: Used by permission of Yuri Malishenko [13].

7.5.2 The Flip Chart

The forgotten flip chart is incredibly versatile. This transportable method provides a guaranteed visual aid. It is most appropriate for smaller settings and groups where you are seeking engagement either in the form of feedback or questions. A basic

Figure 7.7. Whiteboard with magnets and sticky note. Creative use of a sticky note for drawing and colorful magnets for 3D whiteboard illustration. Source: Used by permission of Yuri Malishenko [13].

presentation delivered to a small audience using a well-prepared flip chart could be incredibly dynamic.

In preparation, transfer your story board concept to the flip chart and fill in the details in real-time. Notes can be written in pencil at the top of the page to prompt your next point or as a reminder of critical information. 3M makes a post-it flip chart that can be placed on an easel or stand easily on a table top. Be sure to have good Sharpie markers with the appropriately sized tips and to write in black as the base color and use a primary color like red, blue, or green for emphasis. Place the flip chart on a table or easel where it can be seen easily and stand to the side and away from the chart when speaking.

If using the flip chart in a brainstorming session, be sure there are enough pages available for all possible ideas and notes. Write in print letters rather than script, and if your hand writing is poor, ask a colleague to act as the scribe. When we lead strategic planning meetings, we typically ask someone from the group to act as the scribe so that we can focus our attention on the group's responses and engage all members of the audience. Using the post-it pages, whether in a presentation or brainstorming session, you can place individual pages on the walls in a sequence for the group to see the linear flow of information.

Flip charts could be used to create content rich and quick diagrams visually supporting explanations with a wide variety of creative tools. There are many available materials for working with paper. Professional markers have been specifically designed to be used by graphic facilitators on paper that have no limitation for creativity. For further resource ideas, we recommend Yuri's *Visual Thinking Workbook* or follow his Graphic facilitation posts on social media.

7.6 CALLS TO ACTION

Hopefully, this chapter has done the thing we advise to do: deliver content, show an example, and let it sink in.

Visual aids accompanied with your delivery make all the difference in how the message is received. From slides to flip charts to handouts, we encourage limitless creativity to help enhance engagement, increase persuasive appeal, and imprint memory. Push away from the laptop and open a pad of paper, walk up to the white board, or grab a flip chart. Draw your ideas and employ visual thinking and design before you open the slide template.

REFERENCES

1. Gallo, C. (2014). *Talk Like TED: The 9 Public Speaking Secrets of the World's Top Minds*, 1e. New York: St. Martin's Griffin.
2. Dwyer, F. (2010). Edgar Dale's cone of experience: a quasi-experimental analysis. *Int. J. Instr. Media* 37 (4): 431–437.
3. Fish, K., Mun, J., and A'Jontue, R. (2016). Do visual aids really matter? A comparison of student evaluations before and after embedding visuals into video lectures. *J. Educ.* 13 (1): 194–217.
4. Gibson, J.W., Hodgetts, R.M., and Blackwell, C.W. (1991). Engineering visual aids to enhance oral and written communications. *IPCC 91 Proceedings The Engineered Communication* (30 October–1 November 1991). Orlando, FL, vols. 1 and 2, pp. 194–198.
5. Saleh, T.A. Testing the effectiveness of visual aids in chemical safety training. *J. Chem. Heal. Saf.* 18 (2): 3–8.
6. Dziak, M. (2015). Microsoft PowerPoint. In: *Encyclopedia of Science*. Salem Press.
7. Doumont, J. (2005). The cognitive style of PowerPoint: not all slides are evil. *Tech. Commun.* 52: 64–70.
8. Atkinson, M.(2009). The problem with PowerPoint. *BBC News*. http://news.bbc.co.uk/2/hi/8207849.stm (accessed 27 February 2018).
9. Berk, R.A. (2012). How to create "thriller" PowerPoints® in the classroom! *Innov. High. Educ.* 37 (2): 141–152.
10. Alley, M.P. (2003). *The Craft of Scientific Presentations: Critical Steps to Succeed and Critical Errors to Avoid*, 2e. New York: Springer.
11. Garner, J.K. and Alley, M.P. (2013). How the design of presentation slides affects audience comprehension: a case for the assertion–evidence approach. *Int. J. Eng. Educ.* 29 (6): 1564–1579.
12. Nathans-Kelly, T. and Nicometo, C.G. (2014). *Slide Rules: Design, Build, and Archive Presentations in the Engineering and Technical Fields*, 1e. Wiley.
13. Malishenko, Y. (2016). Whiteboard tips and tricks. *graphicfacilitation*. https://medium.com/graphicfacilitation/whiteboard-tips-and-tricks-92966fdef3fe (accessed 20 February 2018).

8

CREATING CHARISMA

It is not *what* you say but *how* you say it that makes the difference. In this chapter, you will learn how to

- transform your vocal delivery to create a sense of charisma,
- evaluate your body language to ensure you are connecting with audiences, and
- cultivate dynamic delivery techniques.

Rajesh Mishra grew up wanting to be a politician, not an engineer. Growing up in the small town of Raigarh, India, Rajesh believed a career in politics was the way to create solutions for the public. Eventually, Rajesh found himself leaving India for the United States to pursue a master's degree in mechanical engineering at Bradley University. Completing his MA led to his first job in product development at the Ford Motor Company followed by a career at Nissan North America. It was during his time at Nissan that Rajesh realized how his passion for politics might inform his engineering career.

Rajesh's moment of clarity occurred during a meeting between several product managers and engineering teams. He watched as a colleague presented in a style he called "technical overload." To compound this, his colleague's delivery was flat, lacking any dynamism or enthusiasm for the topic. Rajesh watched the product

Engineered to Speak: Helping You Create and Deliver Engaging Technical Presentations, First Edition.
Alexa S. Chilcutt and Adam J. Brooks.

managers' faces as they slowly disengaged and glazed over. What is worse, Rajesh knew he was guilty of delivering presentations in the same way. The thought nagged at him – how many opportunities had he missed?

A few weeks later, Rajesh attended a meeting led by a product manager named Jeff Turk. Rajesh watched Jeff mesmerize a large audience of engineers, managers, and company executives. Rajesh recalled that, "he had this charisma I couldn't quite explain. He spoke with such simplicity and optimism that I was simply spell bound. I realized that if I wanted to get ahead in my career, I needed to figure out how to gain that type of charisma." More impressive was that Jeff's slides had little material on them, usually with only one or two images. Rajesh committed to learning this intangible thing he called *charisma*.

8.1 DYNAMIC DELIVERY

Dynamism, or charisma, means using the voice and the body to create a captivating style of speaking. The pathway to the elusive adjectives of *engaging* or *charismatic* comes down delivery. When we talk to STEM professionals, many are like Rajesh, believing charisma is some elusive trait. In Chapter 2, we dispelled the myth that there are no naturally great speakers. To help you get from where you are as a presenter to where you want to be, you must also learn three elements of delivery: the body, the voice, and practice.

Delivery is the use of the voice and the body to *enhance* a message [1]. The use of body includes facial expressions, gestures, and stance while the voice includes pitch, rate, volume, pausing, and articulation. We have emphasized *enhance* because great delivery builds upon a great message with valuable content.

A dynamic speaker with disorganized and ill-conceived content is still going to be disorganized and ineffective. That is why we have saved the subject of delivery to now. If you do not have your material organized, no amount of charisma will save you. So, if you jumped ahead to this chapter to gain a quick fix for your next business meeting, we encourage you to go back and realize that, like lipstick on a pig, enhancements only work when you have a foundation.

8.2 THE VOICE

Oral communication is an audible medium. Good speakers spend time focusing on how they manipulate sounds in the same manner that a visual artist manipulates paint color. When it comes to delivery, your goal is to paint the best version of yourself.

Many speakers feel that to command more attention or respect, they must adopt a certain tone or manner of speaking that makes people take them seriously. This usually creates a delivery style devoid of emotion and personality. Speakers oftentimes disconnect between who they are in everyday interactions and who they are when the present. The goal is not to become someone altogether different. Authenticity is valued. We begin by examining the voice and have organized vocal delivery into five distinct areas: tone, volume, rate, pausing, and articulation.

8.2.1 Tone

"Bueller? Bueller?" When film actor Ben Stein delivered those lines, he likely had no idea he would be a colloquialism immortalized in American English as a stand in for the typical boring presenter. Stein's lines are not particularly memorable, a math teacher reciting the name of a missing student over and over to a class of disengaged pupils. What makes Stein's lines stick out is his monotone delivery.

Tone refers to how one's voice falls along a musical scale. Tone falls along a vertical axis going from low to high. Every speaker's tone rests somewhere on that scale. Some speakers having a naturally high voice like actress Amy Adams in the film *Enchanted* or the boxer Mike Tyson, while others have a deep resonant tone like actors Lauren Bacall and James Earl Jones.

Your Turn

Take a moment to record yourself saying the following sentence:

This is what my natural speaking voice sounds like, it is the only voice I have and there's no use in hating it.

Now listen back to your recording and consider where your voice falls along a musical scale. Recognizing the tone and pitch of our voice is a first step to creating vocal variety. Monotone refers to speech that stays the same tone during the entirety of speaking. A great example is the Saturday Night Live skit Delicious Dish where cast members Ana Gasteyer and Molly Shannon emulate the monotone soft-spoken style of National Public Radio hosts. Monotone delivery reduces audience engagement and decreases retention of the message [2].

As audiences, we tend to equate changes in pitch that go higher with increased energy, and changes in musical tone that go lower with somberness and gravitas. Variance in tonal dynamics and range by speakers is positively associated with charisma and liveliness [2]. In experiments where audiences were exposed to speakers with limited pitch frequency, listeners found themselves less interested or excited by the speakers. Speakers who used a broader range of high to low sounds were viewed as more passionate about their subject [3].

Using inflection and vocal variety accomplishes more than building audience interest, it also has implications for business. A group of researchers compared the perceptions of speakers versus audiences in a high stakes environment of venture capitalism [4]. The scientists were comparing perceived vocal variety of speakers in contrast to how investors scored their presentation skills. They found that in most cases the entrepreneurs overinflated their perceived vocal variety and overshot the level of perceived charisma. Professionals in technical fields believed that sticking to tightly constructed scripts was preferable to a more extemporaneous speaking style. In this study, the researchers found that entrepreneurs who used vocal variety were viewed as confident, controlled, and passionate. Investors seeking confidence and control were more likely to invest in projects and leaders that exuded those soft qualities in their pitches.

Excellent speaker knows that to build engagement and charisma, they need to vary their tone. Think of inflection as an oral form of italics, you might go up in pitch on the *word* or *idea* that is most important.

Go back to the example we gave you earlier:

> *This is what my natural speaking voice sounds like, it is the only voice I have and there's* **no use** *in hating it.*

Your Turn:

This time use inflection to move your voice higher on the italicized word and lower on the bolded words. Record yourself. Listen to the recording to hear how the meaning of the phrase is enhanced by your manipulation of tone and pitch. We also recommend the following three step program:

1. Be as over the top and ridiculous in your vocal inflection, go up and down many times on erroneous points in your presentation.
2. Go completely monotone, focus on doing the exact opposite of what you did earlier.
3. Try to accomplish the midway point between steps 1 and 2, and you will get a better sense of your use of vocal variety.

8.2.2 Volume

Volume is the most self-explanatory part of vocal delivery. Volume refers to how loud or soft your voice is when speaking. Your volume is a byproduct of both your physical breathing and your confidence. Let us first address a misconception: energy does not equal volume. Some presenters tend to think that if they are trying to convey excitement to an audience, their goal is to be as loud as possible. However, you do need to make sure to adequately fill whatever space you are speaking in with the sound of your voice. Speak for the person at the back of the room. Adjust speaking volume from large spaces to smaller ones. No one wants to be yelled at, but everyone wants to hear you. When you think of volume, we encourage you to consider projection.

Your Turn

Take a deep breath. Inhale. Exhale. If your shoulders moved up and down while you breathed, you might have discovered the key toward creating a good sense of volume.

Good projection of sound comes from the use of air pushing from the diaphragm. The diaphragm is the organ just below your rib cage, and it is ideal for pushing air through your vocal cords with enough force to reach the back of the room. When you breathe through your diaphragm, your stomach should expand and contract with air as you push volume out. This produces a clearer, less breathy sound and ultimately resonates with audiences.

8.2.3 Rate

Speed through the information, and your audience is likely to get left behind. Move through your information at a glacial pace, and you will suck the energy out of the room and get everyone thinking about lunch. According to the National Center for Speech and Voice, the average American speaker talks at a rate of 150 words per minute [5]. In other words, if you were to script an 8–10-minute project update, you are looking at between 1200 and 1400 words.

When we consider the rate of any spoken presentation, consider the listening skills of your audience. Think about where this occasion falls within your Sphere of Influence. In most cases, you are presenting information along with a group of other project managers, team leaders, and other professionals. This means the audience has just gotten accustomed to the sounds of another speaker for 10–15 minutes, and when it is your turn, get them used to the sound of your voice. As such, slow down the first sentence of any presentation you give. Similarly, if you want to convey the feelings of excitement, quickening the pace will likely do the trick.

When you think about the rate of your presentation, consider the role your native language might influence your speed. Studies show that non-native English speakers whose first language originated in the Asia Pacific region are perceived to speak slower than native English speakers [6]. Despite this perception, linguists affirm that differences in speed are better explained due to differences in speaking style than some inherent difference in language [7]. As engineers in a global marketplace, it is advised to use fluctuations in rate to give excitement to your presentations but to be mindful not to assume your native rate will be easily applied to a cross-cultural audience.

8.2.4 Pausing

Think about how much of our day is consumed with endless noise. From the moment you wake to when you lay your head down to sleep, your ears are constantly processing sound signals, such as podcasts, radio signals, and chatter back and forth to our brains. As a result, people can view silence as an experience to be avoided.

When you speak, the use of silence in the form of a pause can be an incredible tool. It can make our ideas stand out and our information absorbed. Pauses signal listeners in a way that more noise cannot.

Consider a painting. Imagine a giant canvas with a solid gray background. At the center of this gray canvas is a bright red circle. No matter what else you try and observe about the painting, the red circle draws your attention. You are certain the artist deliberately painted the circle red and placed it in the center because of the contrast it creates.

While visuals direct the attention of our eyes, a strategic use of silence can direct the attention of our ears through contrast. Think about a typical 10-minute update by the same speaker; we get used to the sounds of their voice and eventually slip into a relaxed state. A pause helps to draw the audience's attention back to the present. This a great device of political candidates, to build toward a specific phrase and pause before saying it. The pause cues the audience that something important is about to be uttered. You can do the same. Want to grab their attention before you say something they need to remember? Pause. Then say it.

Pausing also works to give the audience time to comprehend what you just said. This does not mean that you need to pause after every piece of information you state, but that you need to use pauses to focus attention on allowing the *right* piece of information sink in. In Chapter 6, you learned how to drill down information to the most essential. Likewise, put a pause after what is a particularly insightful piece of content. Earlier in the book, we discussed the concept of cognitive backlog, or the feeling of being overwhelmed by information. Planned pauses are a great way to decrease the likelihood of cognitive backlogging.

In addition to aiding in comprehension, planned pauses can be used to help the audience identify your three main points. Remember in Chapter 5, when we broke down your ideas into main points? Putting a planned pause at each transition helps the audience draw distinction between sections of the presentation. By drawing attention to what is most important, distinguishing between main ideas, and adding a somewhat dramatic quality to your presentation, pauses can punch up any presentation.

Your Turn

The next time you are preparing a presentation, add a pause

- After you present the main thesis or the essential idea of the entire message.
- Between each of your previewed main ideas in your introduction:

    "First we'll discuss the new design specifications our customer requested<PAUSE> then we'll talk about the fiscal impact of our new design change."

- At the end of your main ideas before you move to the transition statement.

8.2.5 Articulation

Articulation is the joining of sounds to make noises that are intelligible – i.e. our ability to speak clearly and for every word to be understood. Sometimes this means working our mouths in ways that ensure an audience can understand what we are saying. One technique is to over emphasize hard consonant sounds.

Your Turn

This next part may be awkward. Trust us and go with it. Read the passage below out loud. Read this out loud and, if you are brave, record it, and listen back to the recording:

To sit in solemn silence in a dull, dark, dock, in a pestilential prison, with a life-long lock, Awaiting the sensation of a short, sharp, shock, from a cheap and chippy chopper on a big black block!

If you recorded yourself, you would hear that, despite your best efforts, some of the sounds you produced were nothing like the words you see on the page. This is a tongue twister Adam practices before he presents as, having been born with a speech impediment, articulation is one of his weaknesses.

Adam used to be incredibly shy, which is hard to believe if you have ever seen him talk in front of a group. As a fifth-grade student, a speech therapist diagnosed him with a severe lisp. The therapist then informed his parents of an ill-conceived exercise wherein they corrected him every time he failed to get his tongue in the proper placement behind the teeth. Every interaction became fodder for criticism and rebuke. Naturally, after a few months of constant correction, he became anxious and incoherent when speaking and almost stopped speaking all together.

Adam overcame his aversion to talking through exposure to speech and debate programs and actively works to articulate so his ideas are comprehended. He does this by making sure the hard "t" and "c" sounds are emphasized so that they travel to the back of any room and bound back. You can also increase this ability by warming up your mouth, so you are putting energy into making sure your pronunciation is effective.

Consider the following for a quick vocal warm up that will get your mouth working:

I am a mother pheasant plucker. I pluck mother pheasants. I am the most pleasant mother pheasant plucker to ever pluck a mother pheasant.

Or you can re-use the material from before:

To sit in solemn silence in a dull, dark, dock, in a pestilential prison, with a life-long lock, Awaiting the sensation of a short, sharp, shock, From a cheap and chippy chopper on a big black block!

Each of these tongue twisters requires excellent articulation, or you will end up saying something offensive. That is why they work.

While we have taken time to discuss each element of vocal delivery, it is important to understand that these elements work together to create variety that reads to audiences as passion and charisma. We encourage you to practice with a focus on each element but realize they work better in concert.

8.3 BODY LANGUAGE

Rajesh Mishra felt part of his charisma challenge stemmed from his background. He noticed that children born and raised in the United States started to present information to their peers in kindergarten. The first time Rajesh was asked to present was in graduate school. He needed training, so he signed up for a two-day public speaking workshop.

The workshop was not the most enjoyable way to spend a weekend. As he stood in the conference room along with 20 or so peers, Rajesh worried he might have made a mistake. When facilitators brought out a large camera and told participants they would be filming participants giving a five-minute presentation, Rajesh considered leaving. Instead, he thought about the two different experiences watching the one colleague commit information overload and the other engaging with the audience and generating enthusiasm. He stuck it out.

Over the course of the next six-hours, Rajesh submitted to being evaluated by experts who noted, among other things, that he stood in one place for long stretches of time. He was forced to listen to audio and note the number of times he changed tone. Fellow participants told him to engage the audience more by asking open-ended questions.

It was a relief to find that simple methods could improve his speaking and presentation skills. He learned the importance of nonverbal communication through use of voice and body. He recalled the time when he dreamt of being a politician. He realized how he thought about politics as a means of bringing value to people, and that his career as an engineer could incorporate the same thing.

Over the next several months, Rajesh actively worked at being as charismatic as possible. He set up after work outings with his native English-speaking colleagues to observe the ways they spoke. He was aware that to understand a group of people best, you needed to learn to understand their communication patterns. Rajesh embraced ownership of professional development.

We covered the voice and the elements of vocal delivery that brings our content to life, now consider the ways you can enhance delivery through the body: eye contact, gestures, and purposeful movement.

8.3.1 Eye Contact

Creating a relationship with your audience is the best way to enhance your persuasive ability. Let us begin with an outward sign of nonverbal connection. The eyes are the window to the souls. This platitude, first expressed by the Greek philosopher Plato, has been repeated so often that you might roll your eyes at the mention of it. Yet, humans are biologically primed to see eye contact as part of connection; we recognize it as either a friendly engagement or a threat [8].

Think about a conversation you have had with a friend or significant other who refuses to make eye contact. You would likely assume the person had some kind of problem with you, and studies illustrate appropriate amounts of sustained eye contact is one of the key ways humans form intimate relationships [9]. When a speaker makes and holds eye contact with you, it increases your level of self-reflection [10]. Audiences perceive higher levels of sophistication, competence, and intelligence when speakers make direct eye contact with them [11]. Studies have shown the ideal amount of eye contact between people is a little over three seconds [12].

When making eye contact with your presentation attendees, use the One Idea/ One Person rule. As you speak, hold eye contact with one person at a time and express one idea. Think about it like speaking a whole sentence to one person, long enough to establish connection, but not too long to where you feel like you are staring the person down. As you complete the first idea, lock eyes with someone else as you communicate the next idea.

Following the One Idea/One-Person rule reduces the anxiety of trying to take in the whole of the group at once. It also has the added benefit of gaining your audience's attention because people do not want someone looking at them while they are looking away. This allows you to gauge the nonverbal responses of your audience and get a sense of their interest. Are they nodding while you are giving this idea? Are they displaying the micro expressions of disgust or resistance?

When you consider eye contact, avoid the temptation to make things easier on yourself. Under no circumstances should you create a fake audience. No looking just above their heads or staring over them toward the back of the room. This might make you feel more comfortable, but the only thing you will accomplish is providing people with an anecdote to discuss at a dinner or happy hour. "What was with the presenter that never looked at us for an hour? That was awkward."

Another disastrous choice is the use of what we call The Sprinkler System. This is a speaker who starts at one side of the audience and scans from left to right and back again, over and over until the audience begins to feel like they are watching the world's slowest tennis match. One idea at a time, delivered directly to one audience member works every time.

8.3.2 Gestures

When we work with speakers individually, gestures tend to be something people fixate on. We once worked with a mechanical engineer who threw his hands up in frustration after watching a video of himself presenting and exclaimed, "I just don't know what to do with my hands!" Many folks feel unsure about what to do with their arms and hands when speaking.

Gestures are a way of adding a visual element that reinforces the crux of what you are trying to say. However, the point is not to choreograph movement for every word. Rather, gestures add nonverbal emphasis and authentic speakers allow their gestures to naturally follow their message.

Just remember two things: keep it open and intentional. Open body language means making gestures that open *out* to your audience rather than into yourself. This includes keeping your palms up and incorporating movement that takes up space.

Avoid closed off language like crossing your arms over your chest. While comfortable, this signals to your audience that you are either insecure or not open to their ideas seeming cold and standoffish. Resist pointing your finger directly at the audience or holding your hand out in the "stop" sign.

The point is not to become a different person, but to present the best version of yourself to an audience. Similarly, you want to seem like you are talking like *you* when you are up in front of an important group. Therefore, if you naturally talk a lot with your hands, keep talking with your hands. Just be conscious of not over doing it. Be natural, but do not look like you are about to take flight.

We are both expressive people and asking us to stop using our hands would be unbearable and make us seem stiff and unexpressive. A good general rule of form, however, is think palms up and hands gesturing in time with your speaking in upward and outward motions. These movements are natural and seen as genuine.

Your Turn

Like Rajesh, you will improve by watching yourself on video. Use your smartphone or video recording device to film yourself talking for one minute about a subject. Consider using the pitch you put together for Chapter 6.

Watch yourself and count the number of distractions you create with your body language. Think of how often you shift your weight back and forth, or if your gestures are closed and accidental. If you created more than three distractions in under a minute, consider the next few steps.

Instead of trying to learn new ways of talking, refine what you already do. With your open body language remember to keep your arm gestures within a box. Image there is a box going from your neck to your waist and keep movements within that framework.

8.3.3 Stance

In the pantheon of poor delivery gestures, stance can be the most egregious. The "I Have to Tinkle" involves hilariously crossing one leg in front of the other and slowly rocking back and forth as your weight shifts from one foot to the other. You are reading this and conjuring up the image in your head of someone who does this. It might be you. Conversely, presenters sometimes keep their legs together and do the "Tinman" where they teeter from side to side.

To get the right stance, hold your body in a way that demonstrates confidence while also being a comfortable position you can hold. This more powerful looking stance incorporates some elements we know about body language and confidence. In her TED Talk, endocrinologist and body language scientists Amy Cuddy reports that when individuals feel more powerful when they use expansive poses [13]. Think Superman or Wonder Woman pose.

When we take up space and hold our bodies in larger ways, we increase our confidence level. In her book *Presence: Bringing Your Boldest Self to Your Biggest Challenges*, Cuddy presents a series of studies that suggest when you make yourself larger in a space, testosterone, the hormone associated with confidence, increases. When you contract your body and fold yourself over in a seat or podium, that confidence hormone decreases [14]. While this can seem silly or out of place, Cuddy's work reminds us that how we carry ourselves changes the way we act and the way we speak.

When you are in front of the room, it is important to engage yourself in what we call a stance that suggestions power and involves the following:

- shoulders back
- hips heels and shoulders in alignment
- evenly distribute your weight
- head held high, eyes looking forward

By positioning yourself in a stable placement, you appear more confident and able to engage whichever audience you choose.

8.3.4 Using the Space

The podium is not your friend. It does not matter whether you are giving your presentation in a board room, classroom, conference room, or an auditorium; the chances are there is a table, lectern, or podium. Speakers see the podium as the space they are supposed to stand behind, after all its likely how you see politicians giving formal public addresses. Other times, presenters use the podium because it creates a safe and comfortable position from which to speak. You can lean on it, and it puts a shield between you and the audience.

The podium is not your friend. Get out from behind it, and center yourself to the group you are speaking to. Distance between the speaker and the audience is related to perceived intimacy. If you want to create a more conversational tone, put yourself closer to the group you are speaking with.

Having planned, purposeful movement helps engage audiences and create distinctions between your major ideas. This involves following the Speakers Diamond (see Figure 8.1). Start at the center of the audience. When you transition between your introduction and point 1, walk to one side of the room and stand in a spot. Repeat this process between points 1 and 2 where you walk to the opposite side. At point 3, return to the center of the room and take a step forward for your conclusion.

Rajesh learned there was nothing more rewarding than connecting with an audience. As he advanced in his career, Rajesh continued to simplify his message and focus on connecting with his audience. Early in his 20-year career, he admitted his

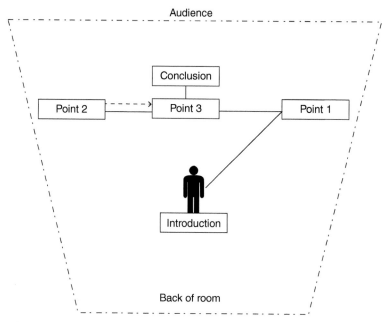

Figure 8.1. Speakers Diamond. This image depicts the pathway a speaker takes while speaking in formal presentations. The speaker begins at the front of the room at the center of their audience, then moves to one side of the room, and then the opposite side for the first and second main points. As they transition to the third point, the speaker returns to the center of the room and takes a step forward for the conclusion.

presentations were near 80–90% technical information and 10% high-level and meaningful presentation skills. Today, as he progresses into upper management at a Fortune 100 construction and mining equipment company, that equation is reversed.

8.4 CALLS TO ACTION

In addition to completing the Your Turn activities throughout this chapter, consider these other calls to action.

- Plan pauses into your presentation to reduce the amount of verbal fillers.
- The next time you sit through a presentation, count the number of dynamic changes to the vocal delivery including: rate, volume, and pitch.

Buckle up. We began this book telling you our goal was to make your ideas more effective and to teach you to make the complex simple and the simple interesting. In Part 1, you gained the recognition and awareness of how communication creates careers. In Part 2, we have shown you how following a process of development makes you a better speaker. In the final part, you will shift from seeing how these tools work to doing them. We take things to the next level to allow you to practice and put these techniques to work alongside other working professionals. We conclude with supplemental materials to keep you on the right path as you continue your journey to success.

REFERENCES

1. Nikitina, A. (2012). Successful public speaking, 105.
2. Hincks, R. and Edlund, J. (2009). Promoting increased pitch variation in oral presentations with transient visual feedback. *Lang. Learn. Technol.* 13 (3): 32–50.
3. Rosenberg, A. and Hirschberg, J. (2005). Acoustic/prosodic and lexical correlates of charismatic speech. *Ninth European Conference on Speech Communication and Technology* (4–8 September 2005). Lisbon, Portugal.
4. Lucas, K., Kerrick, S.A., Haugen, J., and Crider, C.J. (2016). Communicating entrepreneurial passion: personal passion vs. perceived passion in venture pitches. *IEEE Trans. Prof. Commun.* 59 (4): 363–378.
5. Jiahong, Y., Liberman, M., and Cieri, C. (2006). Towards an integrated understanding of speaking rate in conversation. *Proceedings of the Annual Conference of the International Speech Communication Association, INTERSPEECH*, vol. 2, pp. 541–544.
6. Chen, Y. and Robb, M.P. (2004). A study of speaking rate in mandarin speakers of American English. *Asia Pac. J. Speech Lang. Hear* 9 (3): 173–188.
7. Vaane, E. (1982). Subjective estimation of speech rate. *Phonetica* 39 (2–3): 136–149.
8. Myllyneva, A. and Hietanen, J.K. (2015). There is more to eye contact than meets the eye. *Cognition* 134: 100–109.
9. Wesselmann, E.D., Cardoso, F.D., Slater, S., and Williams, K.D. (2012). To be looked at as though air: civil attention matters. *Psychol. Sci.* 23 (2): 166–168.
10. Baltazar, M., Hazem, N., Vilarem, E. et al. (2014). Eye contact elicits bodily self-awareness in human adults. *Cognition* 133 (1): 120–127.
11. Khalid, S., Deska, J.C., and Hugenberg, K. (2016). The eyes are the windows to the mind: direct eye gaze triggers the ascription of others' minds. *Personal. Soc. Psychol. Bull.* 42 (12): 1666–1677.
12. Binetti, N., Harrison, C., Coutrot, A., et al. (2016). Pupil dilation as an index of preferred mutual gaze duration. *R. Soc. Open Sci.* 3 (7): 160086.
13. Cuddy, A. (2012). Your body language may shape who you are. *TED.*
14. Cuddy, A.J.C. (2015). *Presence: Bringing Your Boldest Self to Your Biggest Challenges.* New York, NY: Hachette Book Group.

PART III

COMMIT TO IMPROVEMENT

9

PRACTICE, FEEDBACK, AND ANXIETY REDUCTION TECHNIQUES

This is where the rubber meets the road. If you have gotten this far in the book, now is the time to put it all together. Even after years of speaking to audiences both large and small, we painstakingly practice, seek feedback, and experience moments of self-doubt. We still experience the proverbial butterflies before stepping on a stage and have learned to use that nervous energy to our advantage. The moment you do not experience some nervous energy, you run the risk of delivering a flat presentation. As we wrote in Chapter 2, do not get rid of the butterflies, teach them to fly in formation. In this chapter, you will learn how to

- practice delivering your presentation,
- change the way you give and receive constructive feedback,
- create responses to questions you may be asked following a presentation, and
- employ tactics to reduce anxiety prior to presentations.

You are the next speaker. As you walk to the front of the room, what is going through your mind? In a recent workshop, we asked 60 professionals the same question. Here are a few of their responses:

Engineered to Speak: Helping You Create and Deliver Engaging Technical Presentations, First Edition.
Alexa S. Chilcutt and Adam J. Brooks.
© 2019 by The Institute of Electrical and Electronics Engineers, Inc. Published 2019 by John Wiley & Sons, Inc.

DEATH!!!!

Praying I just don't embarrass myself.

Desperately hoping I can slip out the back door.

Your response may not be this extreme, but any speaking occasion within your Sphere of Influence naturally brings a sense of insecurity. When you place yourself in front of peers or stakeholders, you know your performance will be judged on the merits of the information, personal credibility, and delivery. Whether you are giving a project update, seeking funding, or presenting to a group of community members, the fight or flight adrenaline that accompanies the situation is normal.

In the moments leading to and during speaking, our automatic stress response system produces physiological and behavioral responses that increases blood flow and heart rate and can cause sweating, pupil dilation, and the release of the stress hormone cortisol [1]. This is enough to make anyone avoid speaking situations. We know, however, that making your projects successful and advancing your ideas means never avoiding opportunities to speak. A 2015 study found that public speaking and presentation skills were the top factors that impacted a software engineer's successes or failures [2]. In a 2017 study, researchers examined the effects on the levels of public speaking anxiety by training novice software developers via a virtual reality experience that simulated an auditorium [3]. This approach proved beneficial for the reduction of anxiety, and this type of training is available through other applications, including Virtual Orator© and Speech Center VR©. However, nothing replaces real life practice, self-assessment, and immediate feedback. Public speaking anxiety can seem insurmountable, but there are solutions to manage anxiety and flip the mental paradigm to connect with rather than perform for your audience.

Shifting to the solution for your nerves means understanding that it is normal to experience some form of speaking apprehension and then identifying elements within your control that can eliminate your anxiety. In this chapter, we discuss how to manage anxiety while enhancing skills through preparatory techniques. These include various stages of practice, feedback, tactics for the day of the presentation, and pre-formulating answers to questions. If these steps are adopted, you will not only improve your oral communication skills but help you significantly reduce your level of anxiety associated with speaking.

9.1 PRACTICE MAKES BETTER, NOT PERFECT

If the top 50 TED speakers practiced an average of 200 times for their TED Talks, what does that tell us [4]? We are not advocating that each presentation requires 200 rehearsals, although run-throughs and seeking constructive feedback is necessary. Remember the adage associated with learning of *read it, hear it, see it, say it*? The act of creating your presentation allows you to *read it* and *see it*. Practice allows you to *say it* and *hear it*.

Like any other project you develop, effective communication depends on a series of trials and errors with thoughtful course correction. Speaking skills are best developed through consistent practice. For the speaker, practice achieves improvement in two ways: (i) it strengthens the quality and clarity of your information, and (ii) it allows you to become comfortable with the physiological aspect of delivery. Practice provides information we may not intuitively be aware of, including time, language choice, body language, vocal variety, and visuals. However, endless practice of poor behavior tends to only reinforce what is not working. Therefore, it is vital that your practice contains concrete elements of reflection and evaluation to ensure you are working smarter not harder. Throughout this chapter, we provide several steps for practicing your speaking, thereby reducing the associated anxiety.

Practice Step 1

Let us start with verbally practicing the presentation. It is important to audio record yourself and conduct a self-critique of your material and delivery before seeking feedback. When you record the verbal run through, you may use 3×5 index cards that include main points and other details to remember. This keeps you on track and is typically known as a presentation outline. Unlike the preparation outline from Chapter 5, the presentation outline is a macro-level synthesis of your presentation to use as a quick reference while speaking.

Recording the verbal portion of your run-through provides three important pieces of information: (i) the amount of time it took from start to finish, (ii) the clarity and logical sequence of information, and (iii) the sound of your vocal delivery. Listen to the recording as an objectively as possible and evaluate yourself using the measures below. This evaluation with numerical measures is located in the Supplemental Chapter.

Audio self-evaluation

Time: _____

- ❏ Attention getting device
- ❏ Purpose statement
- ❏ Established credibility
- ❏ Previewed main points
- ❏ Logical sequence of main ideas
- ❏ Transition statements
- ❏ Language is vivid/clear/fluid
- ❏ Voice is used effectively
- ❏ Conclusion is effective
- ❏ Overall effectiveness

Check the time it took from the beginning of the presentation to the point where you introduced your main idea or objective. Was it within the first two minutes? If not, edit in a way that gets you there faster.

Aspects of information or delivery to work change or improve: _____

_____

Look at the recorded time. Remember that going over time is rarely appreciated. If given 20 minutes to present, build in time to take questions from the audience. Create and practice a 10- to 15-minute version of the presentation and have additional information at the ready.

Practice Step 2

Video record a formal run-through using slides and presenting as you would to an actual audience. Once again, you may use note cards to keep you focused and break the habit of looking back at a projected slide. Place the cell phone or camera far enough back to capture you and your slides in the frame. Remember the tip from Chapter 7 on creating slides using dark backgrounds and white text or inversing the color on charts and graphs. This makes for easier viewing. Watch your recording and provide a self-critique, making additional changes in content or delivery. Check the time and adjust if necessary.

Self-Critique Checklist:

When you watch your presentation video, assess your own skills for the following:

Content

    ❑ How well did you gain your audience's attention?
    ❑ Did you have a clear preview statement/thesis?
    ❑ How did you reinforce your credibility?
    ❑ Were there transitions or a clear organization pattern?
    ❑ Was the presentation organized and easy to follow?
    ❑ Did you recap of the most important information as you concluded?
    ❑ Did you end with a strong closing statement or call to action?

Delivery

    ❑ What was the level of energy/dynamism?
    ❑ How did the gestures and nonverbal delivery enhance the presentation?
    ❑ How did you, as the speaker, use your voice and inflection?
    ❑ Were there too many verbal fillers?

To quantify your evaluation or track improvements, see the numerically rated self-critique sheet in the Supplemental Chapter. The next step of practice is to run-through the presentation with a peer, asking for quality feedback. This step is discussed in Section 9.2.1 along with a peer-evaluation checklist.

9.2 GIVING AND RECEIVING FEEDBACK

We see the greatest measure of improvement among colleagues who give and receive honest feedback. When the Harvard Business Review studied the power of emotional intelligence, researchers found that the key for developing effective empathy and relationships between leaders and their employees came down to three things: motivation, practice, and feedback [5, 6].

People often conceive the professional development process as linear; someone tells us information we need to improve, and we immediately apply that information. This may be because the information processing part of our brain is in the neocortex. Areas in this part of our brain tend to respond to one-to-one systems of learning. However, cognitive psychologists tell us that our ability to communicate is located within the limbic system, which requires a different approach [5]. Therefore, your ability to relate and generate emotional intelligence comes from the desire to improve coupled with the ability to practice under conditions of continual impact. Self-awareness begins with the ability to observe and analyze our own behavior. Recording yourself, or having someone else record you, becomes a valuable tool in recognizing how you come across verbally and visually.

A different technique for evaluation involves the Thin Slice approach. This requires recording a full presentation and deeply analyzing a thin slice to generate insights about the overall performance and delivery style of the speaker. Research has suggested that student perceptions of an entire semester's worth of teaching could be based on an analysis of 20-second slices of multiple lectures [7]. There was a high correlation between an observer's rating of the thin slices and the rating of students' evaluations at the end of the semester. By evaluating thin slices of a larger presentation, you allow understanding and "an efficient means to form judgements and predict interpersonal relations from the full-length behavior" [8, p. 19]. If you have a longer presentation, it may be beneficial to watch multiple thin slices to get an understanding of your body language, voice, and presentation style.

Once you have recorded yourself, the analysis can begin. Researchers at the University of South Florida discovered a technique for increasing self-awareness of speakers through observation, analysis, and repetition [9]. The speakers in this study were filmed delivering a short presentation and then, with the help of an assistant, watched the recording and listened for filler words and other disfluencies. As the speakers watched their performance, they raised their hand every time they heard a filler phrase (e.g. *like, um, you know*). The assistants would also raise their hand and would take note of the total number of filler words. After the first round, the speakers gave the presentation a second time and raised their hand while speaking when they used a verbal filler. At the same time, the assistants would raise their hand when the speaker used the phrase without catching it.

This process was repeated two more times until the speaker independently caught every instance. Results suggested that this method drastically reduced the number of verbal fillers used by each speaker [9]. By cultivating self-awareness, speakers can make corrections to their own speech patterns.

9.2.1 Requesting Real Feedback

While it is vital to analyze your own efforts, the previous example highlights that you will not be able to complete this journey toward professional development alone. The thought of unsolicited or negative feedback from someone is not fun. Yet, seeking feedback gives insight to possible changes that may make the presentation better. Asking for feedback is a sign of self-awareness and desired improvement. To do this, you must be open to hearing opinions about your material and performance.

Especially in a technical field, improving your communication depends on opening yourself to the feedback, including cross-disciplinary or nontechnical individuals. You need to ask someone to give you clear, open, and honest feedback. This is the hard part. You will have to negotiate the power dynamics of whatever relationship you are in to indicate that you are open to what others have to say. Let us suggest a few elements that make it a better process.

Choose someone wisely. Your first instinct might be to solicit feedback from someone you interact with every day, such as a spouse or a close colleague. However, the closeness of the relationship might create obstacles in receiving the honest feedback that you require. If you are in a position of authority or well liked in your organization, you might have to be more intentional about asking for feedback. Studies show well-liked individuals receive less specific feedback than individuals perceived as less liked by others [10]. You might ask a mentor or a trusted colleague who may not have a direct role in your day-to-day operations. A spouse might be convenient, but you are more likely to get defensive about the feedback from a loved one than someone you have a working relationship with.

How you ask for feedback will directly impact the feedback you receive. Have a direct conversation in which you accomplish the following:

1. Express that you have realized you want to improve your ability to communicate in the way you speak to and with others.
2. Recognize you are aware of areas that need improvement and that you are taking steps to work on this, which is where you need help from trusted folks.
3. State that you have identified this person as someone you respect and would like to ask for help in guiding your process of self-improvement and give one or two reasons why.
4. Ask for help and give specifics of how you see the process moving forward. This would involve clarifying if this involves watching a video, attending the next speaking event, doing some exercises with you, or just being open about your performance at a recent event where the person had the opportunity to witness your abilities.

Practice Step 3

Run-through the presentation with a friend or colleague as the audience. Ask them to give you feedback about the information, its clarity, organization, and your overall performance. Have the person time you and write notes as you present. Below is an evaluation template that mirrors the outline format provided in Chapter 5. A complete form with numerical evaluations for scoring purposes is also listed in the Supplemental Chapter.

Peer-evaluation

Time: _____

Introduction
- ❑ Gained attention and interest
- ❑ Introduced topic clearly
- ❑ Motivated the audience to listen
- ❑ Established credibility
- ❑ Previewed remainder of speech

Discussion
- ❑ Main points well organized/developed
- ❑ Clearly focused topic
- ❑ Language: vivid, clear, creative
- ❑ Identified with audience
- ❑ Transitions: quality, points well identified

Supporting Materials
- ❑ Quality sources
- ❑ Solid evidence

Conclusion
- ❑ Reinforce purpose
- ❑ Summary clear
- ❑ Sense of closure

Delivery
- ❑ Level of animation/confidence/dynamism
- ❑ Gestures: effective, appropriate
- ❑ Voice clarity, vocal emphasis
- ❑ Extemporaneous style and use of notes

What might be clarified, condensed, or emphasized? _____

_____

What questions are likely from the audience? _____

_____

Practice Step 4

Allow your trusted peer to review the presentation slides with you, one at a time, and offer feedback on content or design. Do not make immediate changes to your slides in response to the feedback. Simply write down the comments and refer to the list later. Your peer may have some good points or ideas and they may not. In the end, you decide what to change or keep.

9.2.2 Providing Feedback to Others

In addition to learning how to accept feedback from others, it is also important to learn how to provide constructive feedback. Providing feedback to others can be uncomfortable. We have previously discussed how public speaking places you in a vulnerable position. Inevitably, you empathize with the speaker and are inclined to give them the benefit of the doubt. People do not want to provide constructive feedback that is taken as criticism. Professional dynamics can complicate delivering feedback; however, this is when it can matter the most.

When Adam worked in corporate public relations early in his career, he wrote a speech for the chief technology officer for a Fortune 50 company. This was in the middle of a major economic downturn where the potential for thousands of temporary and indefinite layoffs loomed. During the filming of the speech, Adam was nervous to give feedback to the Vice President. After the first take, the VP looked over at Adam for feedback. Adam hesitated. The executive looked at him and said, "I hired you to give me feedback, I hired you to solve problems, not avoid creating them." That moment stands out because it was a reminder that providing quality feedback means looking beyond hierarchy and toward solutions.

In case you are asked to give feedback to someone else, there is a way to do this where people walk away with a stronger awareness of their abilities and positive indicators of how they can move forward. The way to do this might seem redundant and obvious, but we are often surprised at how little we see this style being put into practice. The process for giving feedback involves using a sandwich method, guiding questions, and a wrap-up chat.

It may seem somewhat cheesy to think about giving feedback in terms of a Sandwich Method. The method begins with saying something positive about what you just heard, move to something specific and constructive, and return to something positive. The goal is to not place people on the defensive by immediately being critical and suggesting change. Start by supporting them and identify a positive aspect before moving to a suggestion. This increases their receptivity. When providing constructive feedback, keep your language choices active and suggestive. Consider the following strategies, which frame your responses in the positive rather than the negative.

- Use "Try to" instead of "you didn't"
- Use "You might consider" instead of "you failed to...."
- Use "I would have preferred more" instead of "what was missing was"

An example might be – "I appreciate the amount of information you provide on this topic and the strength of your supporting data. It might be helpful to organize the data into meaningful themes for the audience. This way you can focus more on the impact of the data as it relates to them rather than the data itself." Positive, critical, positive will be your friend in feedback meetings for years to come.

Another feedback tactic is to use guided questions. The point is to guide speakers to analyze themselves and create solutions to their issues rather than rely solely on your personal direction. Consider the following strategies:

- What is the goal of this presentation for you?
- What made you want to share this message?
- What are the two things you want people to take away when they leave your presentation?
- How do you feel about that practice you just gave?
- Do you think that the organization pattern you have chosen will allow people to remember the key elements you identified?

As you use guided questions, move from general questions to specific ideas that allow the speaker to narrow focus and come to the same conclusion. This takes longer than simply saying "change your organization, it's confusing." Notice that these questions mirror the questions you asked yourself in Chapter 4. However, the outcomes are stronger because you collaborated on a solution.

Finally, the most important piece of any feedback session is a wrap-up chat. It is crucial that you allow the person receiving feedback to ask clarifying questions and provide a summary of what he or she took away from the practice session. When coaching others, we end every session by asking, "what are two things you are taking from our time together?." This approach allows you and the person receiving feedback to make sure you are on the same page. It also allows the speaker to focus on a few positive results rather than try to retain everything you have gone over. Of course, the most effective feedback sessions involve repeated practice and feedback, this way you are beginning to coordinate those elements of the limbic system vital in constructing powerful presentations.

9.3 MANAGING ANXIETY THROUGH UNCERTAINTY REDUCTION TACTICS

OK, you have verbally practiced and run-through the entire presentation as well as received feedback. We caution you not to over practice but to get to a place of comfort with the material and delivery. The next step toward reducing anxiety is to employ tactics that reduce the uncertainty in your mind.

First, visualize the entire presentation process. Visualization prepares the mind to be in sync with the body. Visualize the space, the audience, hear any possible background noise. If you have ever been asked to speak at a meeting where they are serving a meal, you know it can be distracting; e.g. people clinking silverware, the noises of eating, and small chatter. Visualize you speaking and presenting each slide. Visualize ending with a strong closing statement. Visualize your audience's nonverbal responses. Are they making good eye contact, nodding, smiling, writing notes, and texting? If you create in your mind what the situation will look, sound, and feel like, the less uncertainty surrounds the situation. Less uncertainty leads to reduced anxiety.

If possible, visit the space a day or two prior to your presentation. You may be presenting in the same room your company meets in each week; therefore, you are familiar with the physical space, how the chairs are set-up, and how to work any needed AV equipment. In the past, we have arrived the evening before an event only to realize the seating needed to be rearranged to suit our presentation style or the interactive nature of our workshop. If you are presenting at a conference or any unfamiliar space, you need to see and walk through the space to visualize presenting there. Pull up your presentation and make sure all AV, including sound, is working properly. Once again, reducing areas of uncertainty increases your sense of control. Recall Chapter 7's lesson on the importance of a handout as possible back-up when AV fails.

On the day of, arrive to the space early. This relieves the tension associated with rushing to get there and set-up. You do not need the added adrenaline.

9.3.1 Get Outside of Yourself

Play host. As people are filing in, act like a host. Meet, greet, and make small talk. Whether in a familiar or unfamiliar crowd, a small group of colleagues or larger group of strangers, engaging with as many people as possible loosens you up and breaks down the formal barrier between speaker and audience.

One semester, a student wrote on an instructor evaluation survey that Alexa seemed "aloof and distant" because she stood behind the podium pulling up her lesson as students arrived to class. This caused contemplation. It obviously had not mattered to the student that there were 10-minutes between lectures, or that the prior lecturer never let out on time, meaning that those 250 students were flooding out as the next 250 were flooding in. It was always a challenge to begin on time. Since then however, the focus is on the students as they walk in even if this necessitates multitasking between pulling the AV up and greeting people.

Think of your most formal presentations and how you begin to acclimate the audience. The act of meeting, greeting, and chatting does not just acclimate them; it also helps you get out some of the nervous energy. Standing at the front of the room as people file in and look in your direction only adds to it. Nothing is more awkward than to sit off to the side or stand behind a podium as people are coming in. To wait silently in your seat or behind the podium sets you up for a more formalized beginning and places additional pressure on you.

An additional tactic to employ while you meet and greet is to choose three people you have spoken to designate as friendly faces. These become audience members you can count on to make good eye contact and smile when you look their direction during your presentation. Small encouragements help.

9.3.2 The Audience Is There for You

No one in the audience knows exactly what you are going to say or how you have practiced it. If you lose your place or leave something out, you have the ability to recover. If you have practiced enough that it is a conversation rather than a solo performance, you can adjust with ease. Humility is also a wonderful tool. If you make a blunder, admit it, correct it, and move on. People are not perfect, and they do not expect you to be either. In fact, they would rather a genuine attempt at conveying interesting information than something cold and canned.

When she was 10-years-old, Alexa has a solo dance performance at a summer camp in Atlanta. The camp's patron would also be in the audience. Big stuff for a little girl. Alexa's dance teacher drilled the routine into Alexa's body and head for months. But on the evening of the performance, half-way through the dance, Alexa's mind went blank. She stood in middle of the stage, lights shining, audience looking, and her mind running a hundred miles a second just not recalling the next step. What is a girl to do? After what must have been five seconds, even though it felt longer, Alexa began doing The Funky Chicken, the step she had recently learned. She funky chickened her way right off the stage at which point the patron stood up and applauded. He knew she had blanked but communicated through his applause that it took guts to try and recover rather than run off or break down. That was a great lesson. People are generally rooting for you.

Anxiety associated with speaking can arise from a fear that your audience is sitting there quietly in judgment waiting for you to mess up. In all actuality, your audience is *for* you not against you. Unless this is some type of team competition or you are facing potential defunding, an audience genuinely wants to know what you know, and only cares that you make it interesting and applicable to their needs.

9.4 FIELDING QUESTIONS

You have been laser focused on this presentation. Now, take the blinders off and be objective.

Questions. We want them, yet we dread them. It is like opening Pandora's Box; you never know what you are going to get. Responses to questions gain or lose credibility as well as make or break deals. The higher the stakes, the more you should prepare. Leaving time to take a couple of questions at the end of a presentation is most often beneficial and preparing to answer those questions is as important as the presentation.

If you follow our earlier advice about not fitting 10 pounds of information into a 2-pound bag, then there is plenty of room for questions that will allow you to dig deeper. Think of your upcoming presentation. If you have the outline of your most important points and the items you will be able to cover effectively, there will be aspects of your project the audience wants to know more about.

Your Turn

Write out two questions you will most likely receive.

1._____

2._____

Brainstorming possible questions in advance gives you the opportunity to formulate more efficient and meaningful responses. No doubt we have all listened as a question was asked and the presenter took forever to answer it, never got to the heart of it, or did not answer it at all. Perhaps they went off on a tangent unrelated to the central question. Preparing responses in advance gives you more control of the situation. This does not mean that you know exactly what or how you will be asked something, or that you can prepare and memorize a response. It simply means that you have thought through the most likely questions and given thought to efficient responses.

Audiences appreciate direct responses. You may be so efficient that they ask you to explain something in even greater detail. This also leaves more time for additional questions.

Your Turn

Now, formulate answers to the questions you wrote.

1. _____

   _____

2. _____

   _____

Your Turn

Write two challenging questions, ones you might least want to have to answer.

1. _____

2. _____

How would you respond with your own objectives and desired outcomes in mind? Formulate possible responses.

1. _____

_____

_____

_____

2. _____

_____

_____

_____

_____

9.4.1 Addressing Opposing Viewpoints

If you are pitching, recommending, or selling a project or idea, think about your opposition. Who is your heckler? What are the opposing viewpoints? Who might be at cross-purposes with you?

When presenting at academic conferences, there is always the fear of the expert being in the room or the individual who simply likes to put people on the spot by asking argumentative questions. This is your opportunity. Get into the oppositions' heads to understand their motivations, perspectives, and competing resources.

9.4.2 Stumped?

Phone a friend. In theory, at least. Depending on the question you anticipate, ask a colleague, teammate, leader, or perhaps someone in public relations to review the drafted question and response for feedback. Use this opportunity as a brainstorming session for fielding difficult or controversial questions. The ability to think ahead and craft prototypes of responses will help reduce the level of anxiety associated with the uncertainty of the question portion of the presentation.

Do not know? It is OK! There are ways to handle this with finesse. For example, if you are in a room with other experts, or people who have worked on the same or similar projects, throw the question out to the group. Repeat the person's question and ask the audience if they have additional knowledge that would provide an answer. We have seen this work incredibly well. Typically, there are people with experience that would love to share their ideas, examples, or solutions.

In the worst-case scenarios, you can respond, "I don't know the answer to that off hand, but I will find out and get back to you. Make sure to give me your e-mail address after the presentation." Do not make promises you do not intend to keep and be sure to respond with an answer, even if it is a partial one in a timely manner. Follow this checklist for fielding questions:

- ✓ Encourage them to ask questions (Do not be afraid of silence).
- ✓ Repeat audience questions so the entire audience can appreciate your answer.
- ✓ Keep eye contact with the person who posed the question.

Questions are not comfortable because you cannot control what someone will ask. You know most of the answers, or you would not be the one presenting the information. Remember this is about sharing and connecting, not controlling.

9.5 CALLS TO ACTION

Feeling confident about your presentation from beginning to end is within your control. This is not a one-and-done process where you walk through the essential steps for one presentation, which turns out to be better because of it, and then go back to your data dump and winging it habits of before. Practice may not make it perfect, but there is no doubt that it will make it better. With repeated exposure to speaking, you will develop the skills to take you from good enough to great. This chapter was a guide for preparation and self-development. Embrace the repeatable processes and evaluation sheets located in the final chapter.

Write down the names of two colleagues you could create a feedback loop relationship with and endeavor to help each other in the refinement of your presentation skills.

_____   _____

REFERENCES

1. Kothgassner, O.D., Felnhofer, A., Hlavacs, H. et al. (2016). Salivary cortisol and cardiovascular reactivity to a public speaking task in a virtual and real-life environment. *Comput. Hum. Behav.* 62: 124–135.
2. Peters, L. and Moreno, A.M. (2015). Educating software engineering managers – revisited what software project managers need to know today. *Proc. Int. Conf. Softw. Eng.* 2: 353–359.

3. Nazligul, M.D., Yilmaz, M., Gulec, U. et al. (2017). Overcoming public speaking anxiety of software engineers using virtual reality exposure therapy. *Commun. Comput. Info. Sci.* 748: 191–202.

4. Gallo, C. (2014). *Talk Like TED*. New York: St. Martin's Griffin.

5. Goleman, D. (1999). What makes a leader? … reprinted by permission of Harvard Business Review, 'what makes a leader,' by Daniel Goleman, November-December 1998. Copyright 1998. *Clin. Lab. Manag. Rev.* 13 (3): 123–131.

6. Goleman, D. (2004). What makes a leader? *Harv. Bus. Rev.* 82 (1): 82–91.

7. Ambady, N. and Rosenthal, R. (1992). Thin slices of expressive behaviour – interpersonal consequences. *Psychol. Bull.* 111 (2): 256–274.

8. Ismail, M. (2016). Thin slices of public speaking: a look into speech thin slices and their effectiveness in accurately predicting whole-speech quality. *Commun. Cent. J.* 2: 18–40.

9. Spieler, C. and Miltenberger, R. (2017). Using awareness training to decrease nervous habits during public speaking. *J. Appl. Behav. Anal.* 50 (1): 38–47.

10. Adams, S.M. (2005). Positive affect and feedback-giving behavior. *J. Manag. Psychol.* 20 (1): 24–42.

10

PROFESSIONALLY SPEAKING

In college, I had a professor tell me that the hardest thing you are going to deal with in your career is not going to be physics, or dynamics, or anything academic. It's going to be communication. I thought then, "That makes zero sense to me."
Now twenty years into my career, it makes PERFECT sense.
—Tera Tubbs, Executive Director, Infrastructure and Public Services,
City of Tuscaloosa

* * *

In the previous chapters, we explored elements that make the complex simple and the simple interesting through the way you speak. We have discussed how to expand your Sphere of Influence by making your information meaningful, to enhance your message with visuals, cultivate charisma, and calm anxiety. Throughout this book, you have had the opportunity to learn, try, and do. We encourage you to take what you have learned to the next level. In this chapter, you will learn to

- relate your technical experience to professionals who understand the value of oral communication,
- acquire new methods for expanding your Sphere of Influence, and
- reinforce how speaking leads to professional growth.

Engineered to Speak: Helping You Create and Deliver Engaging Technical Presentations, First Edition.
Alexa S. Chilcutt and Adam J. Brooks.
© 2019 by The Institute of Electrical and Electronics Engineers, Inc. Published 2019 by John Wiley & Sons, Inc.

To help make these lessons tangible, this chapter features professionals who have successfully applied communication skills in their own careers. You will hear from a program executive from NASA who formally worked for the Obama Administration, the CEO of an innovative technology company, an engineer for the Department of Energy, and the Executive Director of Infrastructure and Public Services. In this chapter, we want to highlight what we think are the four pillars of using communication for professional development: embracing your influence, understanding the strategy, putting your audience first, and listening.

10.1 EMBRACING YOUR INFLUENCE

A hallmark of effective communication is the idea that your messages should be well socialized, trying them out with a variety of individuals before you present them more broadly within the organization. In Chapter 3, we focused on the importance of taking your ideas down the hall and encouraged you to expand your influence within and beyond the organization. As we interviewed professionals for this book, we continued to hear stories of how early in their careers they failed to see themselves as fully integrated within the strategy of the larger organization.

Communication is not based entirely around giving formal presentations or writing formal reports but on communicating trust and credibility within a network to get more accomplished. As one professional we spoke to noted, "Sometimes amplifying your message is just about sticking to your talking points in your daily work and letting other people pick up on those talking points without realizing it and starting to beat the drum for you." When technical professionals begin to communicate daily to a variety of stakeholders, they see their ideas get implemented. You may not be the primary decision-maker, but you can embrace your influence by communicating about your projects in ways that are easily identifiable within the larger strategy.

Jenn Gustetic, the Program Executive for Small Business Innovation Research at NASA, and the former assistant director for open innovation at The White House Office of Science and Technology Policy, provides a great example of how a strategic roadshow of her ideas created project success. At the beginning of the Obama administration, there was the Open Government initiative that encouraged federal agencies to be more transparent, participatory, and collaborative. This involved the frightening process of getting large government agencies to open their data and leverage the tools of social media. This was a major change in federal processing. We take for granted the ubiquity of social media today, but, at the time, the idea of sharing government data with the public was a major disruptor. Recalling the situation, Jenn said

> Getting the Department of Transportation (DOT) to release open data, or getting them to have their first social media policy were things I had to think through. I learned how to best advise my client to push that agenda in a department that was generally not open to it. Somethings we tried worked, somethings we tried didn't work. In that role I had

to work with my client to persuade the other key offices within the department of transportation: public affairs, the lawyers, policy office, all the way up to the deputy secretary to get them on board with creating a planning process to do the things that were in the directive. We had to convince them it was worth their time to put in the effort necessary. I had to get them to believe that these were the right things to do and not just do them by writing them in a report but never implementing them.

One of the first actions Jenn's team undertook was preparing a short presentation roadshow. This involved going to different groups in the organization who she believed would be early adopters of the initiative. Jenn moved from her internal to external quadrant from her Sphere of Influence (see Chapter 3). For Jenn, it was important to, "let them see themselves in the idea, then progressively engage the higher and higher resistance folks. By that time, you've already socialized it with enough of their peers that it's becoming something that they can't just say no right off the gate because a lot of people have already said yes." Jenn's ability to allow the resistant voices to air their perspective was crucial to creating the buy-in she needed. She communicated well, listened to the key problems of her audience, and tailored her message to reflect their desires.

Similarly, when Diane Sherman was responsible for technical sales and development at Dow Chemical, she often reached out to different groups and managers about new ideas her team was trying to get off the ground. Diane provided each audience with talking points that helped them present information about changes to their customer. She fit her arguments in relationship to the overall bottom line. Diane said that to be persuasive we must move beyond our favored solution and start the conversation with establishing shared outcomes to help audiences see how proposed changes enable efficient outcomes. By embracing your ability to influence others, you have a much greater chance of advancing your ideas and overall career.

10.2 TAKING THE LEAD

Engineers and technical professionals are often called to share their experience with the public or their broader organization. In these situations, the technical experts often refrain from taking a leadership role in the meeting and instead only answer direct questions related to their specific area. You might be speaking to the news media about a civil project, a guest speaker at an organization luncheon, or on the agenda at a business association meeting. No matter the scenario, it is vital that you take a leadership role in guiding the course of communication.

Early in her career as a civil engineer and as the executive director of infrastructure and public services for a town of 100000 people, Tera Tubbs felt her main duty was to be the expert on project details. After all, she was the lead on several projects with major impacts on city neighborhoods and tax dollars. Before her first public meeting, Tera drilled herself on the facts. "I felt like I had to know everything about the project. I would get so nervous making sure I knew everything from the technical side." She spent time asking herself, "Why did they decide to do that here, and why did we make it that size?" She was scared not to be able to answer questions from reporters and citizens on camera.

The next day, after hours of preparation and little sleep, she arrived at the meeting ready to fire away at the first technical question. Only, no one asked a single technical question. Tera was shocked. She realized her approach had failed to account for the specific needs of her audience. She was focused on the problem from *her* perspective instead of the perspective of those she was speaking to. Remember in Chapter 2 when we dispelled the myth of the Sage on Stage? Tera realized that people do not really care about the technical details; they care about the effects the project has on them like, "How long will my road be closed?" She started to ask herself, "What does the audience really want to hear?" and said, "you give them that and only that. You are the expert, otherwise they wouldn't be asking you to speak. When I was able to recognize this my career really took off for me."

Tera encourages professionals to pay attention to their audience's interests. "I don't mean pay attention to the content but pay attention to *how* they are getting the message across." She suggests practicing reflective techniques after notably bad or good meetings. Ask yourself. "What was it that made the meeting go so well?" or "How did that go so wrong?" Tera remembers the day she learned the importance of leading a meeting:

> I remember very early in my career, I went to my first community meeting as the guest speaker. There was a certain building in town, which had been hit by a car several times. Every time the building was hit, the driver was intoxicated, but for some reason this was a problem the city needed to fix. You know as well as I do, that is not feasible, but I really got beat up and beat up over the fact the city had not done anything to prevent the accidents. The more I did not take control of the meeting, the more it turned into a let us pick on Tera affair. It was a terrible experience. When it was over, I told myself I will never let myself go through anything like that again. I have learned if you are in front of an audience and you are in the position to speak and lead the meeting, then you need to *lead the meeting*. You cannot let someone else take charge of the meeting. If you do, you are guaranteed not to be successful.

For Tera, failing to embrace her influence meant she had limited her capacity for leadership to communicating only technical details. Remember when we discussed the fear of communication is not a fear of being bad, but of being good? Tera's example reminds us that when we take the opportunity to lead with meaning by communicating clearly and making the complex simple for people, not only do we increase our Sphere of Influence, but we save ourselves from complicated interactions.

10.3 BEING PART OF THE STRATEGY

An issue with learning effective oral communication is both a persistent stigma of the awkward engineer, and a lack of time and focus placed on curricular support of strategic communication skills. As we discussed Chapter 2, a key part of improving your communication abilities comes from embracing the parts of engineering education that allow you to excel. "Engineers are not trained how to ask questions; they are trained in how to SOLVE questions," said Jenn Gustetic. She described a common scenario in engineering education where individuals are given a problem set, a known and an unknown, an A to be solved eventually arriving at B. Engineers are exemplary at solving from A to B. What they are terrible at is defining A. When you are communicating, you must look at the strategic opportunity and ensure you are first clearly defining the problem. She gave the following example:

> They might be working on a technology program and identify there's a system issue with a screw that didn't fit. Well [you have to ask] is that because that one screw was made badly or are the whole batch of screws bad. They might identify a problem, and then proactively try to solve that problem, but it's usually a specific problem they've already thought about solving. They're not asking larger scale questions like should it even be a screw in the first place. Engineers are trained and wired to defend their solution, which means it can be hard for them to get out of their own viewpoint that this solution is great. "Look at this. I've got this rock, this rock is going to solve all these problems. This rock is great." when no one even wants the rock. It is valuable, but it's not solving a bigger strategic problem.

Jenn's example is a reminder that strategic communication involves understanding the relationship between the individual and the organization, the team and its customers, and the product and its people. The goal is to locate yourself inside that matrix to determine what people need and how what you are offering will accomplish their goals. "Your whole career cannot be about selling your personal pet rock," said Jenn, "because that rock might never take shape anywhere, and you might be missing the thing sitting on your desk that your stakeholders are begging for." It is vital that when you are preparing your own presentation, you make sure you are addressing the real problems, and not merely presenting your solutions. Being part of the strategic thinking process enables your career growth.

10.4 BREAKING OUT OF BAD

Aside from the lack of communication training, sometimes those within the field encourage other professionals to accept the status quo of delivering poor to average presentations. Remember when we told you most people are not afraid of being bad? We spoke to one CEO who had his own realization of what happens when we fail to communicate.

Noah Zandan is the CEO and Co-Founder of Quantified Communications. His company provides assessment and professional development for CEOs by pairing quantifiable language and oral performance characteristics with leadership principles. Noah started his company after a series of realizations about many professionals' lack of skill in speaking and presenting and saw this play out in high stakes situations.

Noah was a summer intern at Lehman Brothers in New York City and was tasked with helping promote a particularly promising IPO for a lucrative real estate company. He and his colleagues spent hours putting together the investment pitch and headed off on a private jet for a 30-city tour. "We carefully sweated all the details of putting the pitch together, the model we created, and we prepared vigorously for the Q and A. I remember I was watching how everyone was preparing. They were so dry in the way they communicated, talking totally monotone. All the guidance the speakers were getting was in direct conflict with the communication theory I had gotten in just one class in college."

Noah completed a public speaking course as a Dartmouth undergraduate, and watching his bosses violate everything he had been taught was frustrating. They told speakers to "memorize your script, do not speak extemporaneously; it's best if you get every word exactly right when speaking to financial audiences." Millions of dollars were on the line and he became baffled by the inability to tell the story of the company in ways exciting enough to make people excited about the IPO. It was seeing this moment repeat itself throughout his early Wall Street career that led Noah to recognize that quantitative people rarely focused on cultivating qualitative skills.

As he moved from Lehman Brothers to another job on Wall Street, and then to a private equity firm in LA, he received endless training opportunities in Excel, accounting, and financial modeling, but he was never given a chance to develop his speaking skills. He understood success in his financial roles relied on "how you build relationships with investors and management teams, and can you go present information correctly in a way that is engaging" but there was "zero guidance" in how to do that. It was during his MBA program at Northwestern's Kellogg School of Management that he began having discussions with an expert in reputation management about creating a data analysis program to develop speaking skills for data driven people.

Zandan's company Quantified Communications, asks leaders to submit videos of a speech or presentation that is then run through algorithms to analyze the communication for variables like engagement, persuasion, and confidence. The video is then compared against rated stimuli from panels of experts to produce an overall score that is given to leaders.

Noah has taken professional development to the next level. He recognized delivering feedback is more accepted if you can integrate objective measures, "If you bring objective pieces of evidence to these quantitative people you are more likely to get them to listen. People are more comfortable making decisions based off data." With this type of technology available, there really is no excuse to continue to be bad.

10.5 READING THE SIGNS

We have already discussed the power of understanding your audience. And for several professional engineers, understanding this power became a key to career success. However, it is not just about knowing what stakeholders want, but about reading them in the situation. Tera Tubbs recalled a time early in her career when she found it, "easy to stand up in front of a group of people and say we've coordinated with ALDOT and we're going to install 24-inch RCP and fund the project from the RFFI. I would look at the audience and see blank looks on the faces of the people to whom I was talking. No one knew what it meant." Tera assumed her audience knew what the Reserve Fund for Future Improvements was because her team worked with that fund daily. Fortunately, she was able to read her room and realized her message was not making the kind of impact she wanted.

Understanding your audience means looking for signs of body language that reveal a lack of understanding. In her role at NASA, Jenn Gustetic believes, "You've got to learn to read your boss and the room. There's a lot of bad assumptions about what the audience knows already." Technical professionals need to learn to recognize the signs of good or poor understanding in the person you are reaching out to. She said, "A quick way to frustrate a leader is to belabor a point. Why spend 15 minutes on something you could spend 2 minutes on? Engineers suffer from thinking the process is more interesting than the outcome, and the data that you used to support your finding is so interesting that you want to dive into the data."

To correct this, lead with meaning and stay on point with what matters to that specific group. Jenn calls this "putting the end at the beginning." She uses this process as she plans out communication or status updates, asking herself, "are you focusing enough on the things that help drive action, or do you have other stuff in there to distract them? Is the data I'm discussing leading toward the take away I want?" For status updates, Jenn recommends engineers begin from a simple outline that mirrors the STAR method we discussed in Chapter 6:

- One sentence about the problem you are addressing (Situation)
- One sentence about the deadline (Task)
- What were you trying to accomplish? (Action)
- What were the outcomes in the most recent tests (Result)

Jess recalled with frustration, "I can't tell you how many times I've been presented with a solution and I have to ask, or worse, wonder what the problem was in the first place. It's like here's this great solution but I have no idea what you are actually addressing."

Addressing the topic in a succinct and functional manner will go a long way in keeping your audience engaged. However, continue to read the signs. Are they looking like they are listening? Are you receiving eye contact and looks of understanding? If people look confused, ask them if something needs clarification. Include an example that makes the concept concrete, or begin to use more easily understood terms. Are you seeing boredom on the faces of your audience? If so, move on. Be willing to change it up even in the moment to reengage your audience. Connection is key.

10.6 LISTENING

Professionals reinforce the power of listening. What is heard is more important than what is said.

For Jenn Gustetic, communication is "not just what you meant to say but what was heard." Jenn runs a process of annual solicitation that sees over 1500 proposals in a year. Her team down-selects those to the top 400 or so to follow up for funding for the highest priority R&D that aligns with NASA's mission. With thousands of companies applying, and thousands more involved in the review process, she leads a system that is over constrained and where you cannot make everyone happy. Instead she is, "constantly in listening mode to understand what the real systemic issues are for users and not just one person's pet peeve and prioritizing what to respond to. Without the communication skills I've developed I wouldn't be able to make the program improvements I'm making." For professionals like Jenn, the responsibility for an outcome requires a sense of whether your message has been adequately heard before it can be implemented.

Jenn advocates for professionals to drive for understanding in every conversation they have, and encourages others to engage in clarifying questions to ensure they truly get at the key issues of a situation. She wishes this was industry standard practice, but unfortunately it is not. This means those who can capitalize on this skill gain tremendous advantage over their peers. Throughout the course of her career, she has developed a means of informally assessing members of her team based on team members ability to listen:

> If I'm working with you, or you are on my team, or you are a collaborator on another program I'm working on, and I ask you to do something once and you give me exactly what I ask for, I'm like: Oh my god you're a unicorn! You do not exist, I will never let you go. Because you are a good listener and you know how to drive action from that listening.

> On the other hand, I fully expect that even top performers might need me to restate the same thing a few times or ask clarifying questions. Maybe they don't know me as well, or maybe I didn't state it well and they need me to re-state the requirements. I find it reasonable for people to ask 1-3 times for clarification, and I'll still put you on the high performers list. If I'm having to repeat myself 5 times I start to get to the point where I question if the benefit of what I get on the back end worth the amount of time I spend clarifying.

At the point where Jenn repeats herself seven or more times, she begins to place people in the category where she questions their ability to listen and reconsiders their professional efficacy.

According to Jenn, the ability to ask clarifying questions is somewhat impeded by professional stigmas and the desire for engineers to always appear that they are the most knowledgeable. Engineers do not want to seem like they do not understand, which is why they have a hard time asking the right questions.

With engineers there's this drag to precision, we get so focused on literally interpreting the task and not taking our head's back out further and thinking, what is the strategic opportunity here. What beyond the specific ask should I be thinking about in communicating the broader impacts of this. We can get very pigeonholed.

Break the stigma that it is not ok to raise your hands and ask the right questions. If you have a question, it can generally be assumed someone else does as well. In fact, according to Gustetic, "There's courage in asking clarification questions, and it's something I take note of in my teams".

For Tera Tubbs, listening is also the quickest way you learn how to communicate. "If you go into a room and listen to the questions that are asked you can start to really discern what is important to that audience and then you give them the information that they really need," she said. "I think a lot of people forget to listen they just look at their project and they assume that something is important because it's important to them." She sees a problem when engineers go into spaces with the mentality that, "these are the 10 things I'm going to talk about. When you finish your list of 10 things you might not have answered anyone's questions and what's worse is they are now upset with you for wasting their time."

10.7 FINISHING STRONG

Making your expertise known and your ideas accessible means improving your oral communication skills. For practicing engineers and technical professionals, sharing ideas with a variety of audiences means self-awareness and a desire to grow your Sphere of Influence, asking the right questions, applying principles of organization, putting your audience first, and making the commitment to create charisma. Throughout this book, you have learned to recognize how communication operates in your life, seen how to take tangible techniques for improvement, and done the work you need to do to take things to the next level.

Your work is not yet finished. We know that this is a process of growth and development, of support from inside and outside of the engineering team or organization. For this purpose, we created this book as an ongoing resource, something you can turn to the next time you are asked to speak, or the next time you have a big idea you need to take down the hall. To help you along this journey, we have provided supplemental material in the next section. Within the proceeding pages you will find

- the step-by-step process for developing a presentation,
- practice and feedback evaluation forms for repeated use and measurable outcomes,
- additional templates and visual formatting suggestions, and
- training curricula for corporate or academic implementation.

Learning to make the complex simple and the simple interesting is not about overcoming anxiety or turning weaknesses into strengths. It is about ensuring that great ideas flourish. Excellent oral communication for engineers means increasing efficiency and return on investment by allowing that technical talent to thrive.

APPENDICES

SUPPLEMENTAL RESOURCES

This section includes the active portions of each chapter and a 10-session training/ curriculum in oral communication. We walk you through the process of creating a presentation beginning with self-assessment of personal communication strengths and weaknesses to outlining a pitch or presentation and delivering it with confidence.

The curriculum was developed for a 10-session Research Experience for Undergraduates program. It can be taken apart and easily incorporated into training and professional development in organizational settings or existing technical coursework.

- Appendix A: Self-Assessments from Chapter 1
- Appendix B: Sphere of Influence/Active listening from Chapter 3
- Appendix C: Asking the Questions from Chapter 4
- Appendix D: Organizing and Outline Your Presentation from Chapter 5
- Appendix E: Perfecting Your Pitch from Chapter 6
- Appendix F: Visualizing Your Message from Chapter 7
- Appendix G: Creating Charisma from Chapter 8
- Appendix H: Practice, Feedback, and Anxiety Reduction Techniques from Chapter 9
- Appendix I: Ten Session Communication Curriculum

Engineered to Speak: Helping You Create and Deliver Engaging Technical Presentations, First Edition.
Alexa S. Chilcutt and Adam J. Brooks.
© 2019 by The Institute of Electrical and Electronics Engineers, Inc. Published 2019 by John Wiley & Sons, Inc.

APPENDIX A: SELF-ASSESSMENTS FROM CHAPTER 1

Communication Assessment Questions

For each of the below statements, assess your personal ability using the scale from 0 (no ability) to 7 (great) as they apply to your work environment or professional situations.

0 – No Ability   7 – Great!
         0-------1--------2-------3-------4-------5-------6-------7

1. Feeling confident to share your ideas verbally.   _____
2. Communicate verbally with those in your team.   _____
3. Communicate verbally to those in management or leadership   _____
    positions.
4. Share ideas in a formal group setting (i.e. meeting).   _____
5. Advocate for a specific action or point of view.   _____
6. Understanding other's motivations/perspective.   _____
7. Listening to gain understanding.   _____
8. Reading someone's body language.   _____
9. Awareness of the nonverbal (body language) cues you display   _____
    during an interaction.
10. Self-awareness of vocal variety (volume, pitch, inflection) used _____
    during interactions.

Level of Anxiety in Public Speaking Situations

1. You have been asked to speak to a group of professional peers or to deliver a presentation to potential clients/investors. Rate the level of anxiety you experience prior to speaking?

    0 – No sweat!   10 – I'd rather be mauled by a bear!
       0-------1--------2-------3-------4-------5-------6-------7------8------9------10

If you placed yourself at the lower end of the scale, you are less anxious about speaking to a group. That is great. There are a great number of other elements of the speaking situation and ways to improve your speaking and presentation skills. If you placed yourself at the higher end of the scale, meaning you are more anxious or would rather be mauled by a bear than speak to a group, we can help! Not only will you walk through the methodology to make the presentation good, but Chapter 9 will provide you with methods to gain control over any speaking situation and tactics for reducing stress associated with public speaking.

2. Which part of the speaking/presenting process do you find most challenging?

   ❑ Crafting effective messages that resonate with my audience.
   ❑ Cultivating dynamic delivery (use of nonverbal gestures, voice, movement).
   ❑ Creating a strong close or "ask."
   ❑ Calming anxiety.

   Identifying which areas of message creation or delivery you find challenging will help you focus on certain aspects of improvement. However, we will be asking you at a certain point to get others' feedback of your presentations or shorter pitches to gain an objective perspective. We do this regularly, subjecting ourselves to constructive feedback for continuous improvement. There have been times where we thought our message or structure was clear, but someone's feedback told us otherwise. Be open not just to learning, but to professional growth.

3. How likely are you to apply creative ideas and tactics to make a presentation engaging or memorable?

   0 – Not Likely   10 – Highly Likely
   0-------1--------2-------3-------4-------5-------6-------7------8------9------10

4. What thoughts go through your mind immediately before speaking?

   _____

5. Describe your presentation challenges. Discuss areas of speaking or presentation skills you feel the need or desire to improve.

   _____

   _____

   _____

APPENDIX B: SPHERE OF INFLUENCE/ACTIVE LISTENING FROM CHAPTER 3

Sphere of Influence

Your Sphere of Influence consists of a Core surrounded by three layered circles that are dissected into four quadrants and represent all possible communication opportunities for the technical professional. See Chapter 3 for a full description of the model.

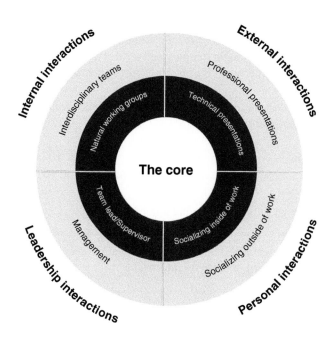

Take time to create your own Sphere of Influence using the model below. In each quadrant, identify opportunities to communicate with individuals or groups according to the sector. This activity will allow you to recognize potential areas of professional communication growth and development.

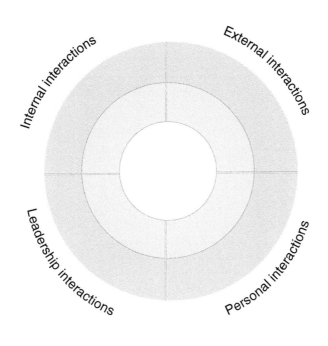

Active Listening

List the variety of individuals you must share your ideas with to create successful outcomes.

_____   _____

_____   _____

_____   _____

Four Active Listening Tactics

Mentally and physically placing yourself in a receptive and responsive mode. Listening is an intelligence gathering activity. If you can listen for meaning, intention, and need, you can decide what parts of your knowledge fill an information gap and decrease ambiguity for others. When interacting with colleagues and stakeholders, listening should be a priority that guides how you share information and craft responses. Remember to paraphrase, prime, express understanding, and use nonverbal cues to tune in and build rapport, all helping you target their need and formulate appropriate responses.

Have you used active listening methods? If so, what responses or results have you noticed?

_____

_____

APPENDIX C: ASKING THE QUESTIONS FROM CHAPTER 4

When tasked with presenting information, take out a piece of paper and write the answers to the following:

1. Who am I speaking to? Who needs the information?
2. What is the purpose of my presentation?
3. What is the desired outcome?
4. What information matters most?
5. Why should my audience care?
6. When am I speaking?
7. Where am I speaking?
8. How will I present? What types of visual aids would supplement my message?

Audience Analysis

1. Describe the audience. Write a list of characteristics that apply to them.

   _____   _____

   _____   _____

2. Are they captive or voluntary?
3. Technical, non-technical, or a mix?
4. What is the desired outcome? What do you want them to *do* with the information?

   _____

   _____

5. What is their degree of knowledge about the subject or project?

   _____

6. What are possible attitudes or pre-disposed biases they may hold either for or against your presentation topic?

   _____

   _____

APPENDIX D: ORGANIZING AND OUTLINE YOUR PRESENTATION FROM CHAPTER 5

In Chapter 2, we asked you to think of an upcoming presentation of your own and answer the foundational questions, *Who, What, Where, When, Why and How*. Now to develop an outline following the template:

Topic:
General Purpose: To inform/To persuade
Specific Purpose: (What are you trying to accomplish?)

Introduction
- Attention-getting device (AGD)
- Reveal topic
- Credibility statement
- Preview main points

Informative Body of Presentation
- Main point 1
  - Sub-points (evidence, explanation, examples)
- Main point 2
  - Sub-points
- Main Point 3
  - Sub-points

Transition Statement between each main point.

OR

Persuasive Body of Presentation
- Three main points [Problem–Cause–Solution or Cause–Effect–Solution]
- Develop each main point with two to three sub-points and identify with audience
- Transition Statement between each main point

Conclusion
- Restate topic
- Review main points
- Close with Return to AGD or Clear next steps/Call to action

Topic: _____

General Purpose: *Circle one*:   To inform.   To persuade.

Purpose Statement: To inform the audience about _____. *Or*

To persuade the audience to _____.

Introduction:
Attention Getting Device:

_____

_____

Reveal Topic:

_____

_____

_____

Credibility Statement/Source:

_____

_____

Preview of Main Points:

_____

_____

Transition statement to first main point.

_____

_____

Body:

I. First Main Point

_____

_____

A. Sub-point – supporting data or information

_____

_____

B. Sub-point – supporting data or information

_____

_____

C. Sub-point – supporting data or information

_____

_____

Transition statement:

_____

_____

II. Second Main Point

_____

_____

A. Sub-point – supporting data or information

_____

_____

B. Sub-point – supporting data or information

_____

_____

C. Sub-point – supporting data or information

_____

_____

Transition statement:

_____

_____

III. Third Main Point

_____

_____

A. Sub-point – supporting data or information

_____

_____

B. Sub-point – supporting data or information

_____

_____

C. Sub-point – supporting data or information

_____

_____

Transition statement:

_____

_____

Conclusion:

Reiterate topic/Purpose:

_____

_____

Review main points:

_____

_____

AGD or Call to Action:

_____

_____

Fielding Questions: Create at least two possible questions and your responses.

_____

_____

_____

_____

_____

_____

_____

_____

_____

_____

_____

APPENDIX E: PERFECTING YOUR PITCH FROM CHAPTER 6

Creating Tangible Narrative Examples

STAR Method: Situation, Task, Action, Result

Build compelling narratives of tangible results. The STAR method takes you from something, through something, to something.

Situation: What was the initial situation?

_____

_____

_____

_____

Task(s): What task(s) were you assigned or challenged with?

_____

_____

_____

_____

Action: What actions did you take to complete the task(s) successfully?

_____

_____

_____

_____

Result: What were the results? How was the situation improved or resolved?

_____

_____

_____

_____

Pitch Framework

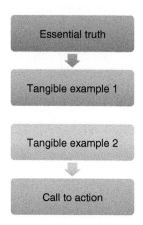

This the formula for an effective pitch is easily adapted to fit within given circumstances. Use it to convey personal value or specific project arguments to others. It helps you develop a set of talking points for a variety of speaking occasions.

- Essential Truth – Describe what you do/bring to the table in a short statement (nine words or less).

  Adam and Alexa's: We help others articulate the value of ideas effectively.

  _____

  _____

- Tangible Example 1

  Ours: We worked with a corporate operations management team who wanted to "gain a seat at the table" to be included in initial conversations about project development. We worked with the group to articulate the value of OM to structural development and project success. They practiced responding to typical scenarios with the intent to change the stakeholder's perceptions of their roles and include them in initial conversations. They have since become key players in leadership meetings and have facilitated streamlined processes and successful implementation of innovative ideas.

  _____

  _____

  _____

  _____

  _____

- Tangible Example 2 (Optional)

_____

_____

_____

_____

- Call to Action

_____

_____

_____

_____

APPENDIX F: VISUALIZING YOUR MESSAGE FROM CHAPTER 7

Storyboard: Place one concept in each box with associated images, equation, diagram, or object.

What type(s) of visual aid(s) would be

most appropriate for key concepts?

_____

_____

Types of Visual Aids and General Uses

Listed in the left column of this table are the most common types of visual aids. Beside each type is a description of the ways presented.

Visual Aid	Uses
Slides	*Projection of slides containing images, text, color, and embedded links, usually in a sequence following the material.*
Overhead projector	Demonstrate work solving problems, illustrating, etc. in real-time using transparencies. May have pre-prepared transparencies with partial work.
White board	Writing out points, showing work solving problems, illustrating in real-time. Marker size depends on size of room.
Flip chart	Making lists, brainstorming, showing work in small groups.
AV – Video / Audio	Show recording of movement or relevant topic video or listen to audio-only of specific sound.
Physical object	*Exhibit physical item for audience to see, touch, feel, and interact with subject matter.*
Demonstration	Show or display process through physical motions or manipulation of an object.

Dos and Don'ts of Slide Construction

The left column of below table contains a list of constructive considerations, Dos, when crafting your slides. The right contains a list of negative presenter habits that create a busy or confusing slide, the Don'ts.

Do Use	Do not Use
Claim or assertion at top in sentence format	Text-heavy slides including overuse of bullets, paragraph format
Clear, high quality images and graphics representing key concepts	Busy diagrams or graphs with too many labels or competing data
A key equation that summarizes point	Multiple equations on single slide
San serif font	Multiple types or stylized fonts
Dark background with white text	Neon backgrounds or pastel color fonts
Larger fonts (≥ 24)	Small font (≤ 20)
Consistent background and colors throughout	Text or slide animations for emphasis
Attributions of images if not original	Too many images on a single slide
Embed videos / include audio if sound is applicable	Videos linked to outside source

Assertion-Evidence Slide Design

1. Build each slide based on one key message as an assertion (full sentence).
2. Support that message visually.

Place your assertion or declaration as a complete sentence at the top of slide inside header box. Below assertion, place the visual evidence to support or explain your claim. This may include photographs, data in a simple format, or diagrammatic processes. More found at https://www.assertion-evidence.com

APPENDIX G: CREATING CHARISMA FROM CHAPTER 8
Vocal Delivery

Tone: Tone refers to how one's voice falls along a musical scale. Tone falls along a vertical axis going from low to high. Every speaker's tone rests somewhere on that scale. Some speakers having a naturally high voice like actress Amy Adams in the film *Enchanted* or the boxer Mike Tyson, while others have a deep resonant tone like actors Lauren Bacall and James Earl Jones.

Take a moment to record yourself saying the following sentence:

> This is what my natural speaking voice sounds like, it is the only voice I have and there's no use in hating it.

Now listen back to your recording and consider where your voice falls along a musical scale. Recognizing the tone and pitch of our voice is a first step to creating vocal variety.

Volume: Volume is the most self-explanatory part of vocal delivery. Volume refers to how loud or soft your voice is when speaking. Your volume is a byproduct of both your physical breathing and your confidence.

Take a deep breath. Inhale. Exhale. If your shoulders moved up and down while you breathed, you might have discovered the key toward creating a good sense of volume.

Good projection of sound comes from the use of air pushing from the diaphragm. The diaphragm is the organ just below your rib cage, and it is ideal for pushing air through your vocal cords with enough force to reach the back of the room. When you breathe through your diaphragm, your stomach should expand and contract with air as you push volume out. This produces a clearer, less breathy sound and ultimately resonates with audiences.

Rate: Rate is the speed with which you speak. Ask others for feedback concerning your rate of speech and ask if you are speaking too slow or quickly. Goldilocks. Think of your audience and the time they need to process.

Pausing: To help distinguish important points and redirect your audience's attention, use pauses strategically. The next time you are preparing a presentation, add a pause:

- After you present the main thesis or the essential idea of the entire message.
- Between each of your previewed main ideas in your introduction:
  > *"First we'll discuss the new design specifications our customer requested<PAUSE> then we'll talk about the fiscal impact of our new design change."*
- At the end of your main ideas before you move to the transition statement.

Articulation: Articulation refers to the way sounds join to make noises that are intelligible to an audience. Meaning articulation is our ability to speak clearly and for every word being said to be understood. Read the passage below out loud. Read this out loud and, if you are brave, record it and listen back to the recording:

> *To sit in solemn silence in a dull, dark, dock, in a pestilential prison, with a life-long lock, Awaiting the sensation of a short, sharp, shock, from a cheap and chippy chopper on a big black block!*

For a quick vocal warm up that will get your mouth working, and get you stepping a little outside of yourself, consider the following:

> *I am a mother pheasant plucker. I pluck mother pheasants. I am the most pleasant mother pheasant plucker to ever pluck a mother pheasant.*

Each of these tongue twisters require excellent articulation or you will end up saying something very offensive. That is why they work and are fun as well.

Body Language

Eye Contact: One Idea/One Person: As you speak hold eye contact with one person at a time and express one idea. Think about it like speaking a whole sentence to one person, long enough to establish connection, but not too long to where you feel like you are staring the person down. As you complete the first idea jump around the room to lock eyes with someone else as you communicate the next idea.

Gestures: A good general rule of form is think palms up, hands gesturing in time with your speaking in upward and outward motions. These movements are natural and seen as genuine.

Your Turn

Like Rajesh, you'll improve by watching yourself on video. Use your smartphone or video recording device to film yourself talking for one minute about a subject. Consider using the pitch you put together for Chapter 6.

Watch yourself and count the number of distractions you create with your body language. Think of how often you shift your weight back and forth, or if your gestures are closed and accidental. If you created more than three distractions in under a minute, consider the next few steps.

Instead of trying to learn new ways of talking, refine what you already do. With your open body language remember to keep your arm gestures within a box. Image there is a box going from your neck to your waist and keep movements within that framework.

Stance: To get the right stance, hold your body in a way that demonstrates confidence while also being a comfortable position you can hold. When you are in front of the room, it is important to engage yourself in what we call a "power stance" which involves the following:

- Shoulders back
- Hips heels and shoulders in alignment
- Evenly distribute your weight
- Head held high, eyes looking forward

By positioning yourself in a stable placement, this allows you to appear more confident and able to engage whichever audience you choose.

Using the Space: If you do not remember anything else, **the podium is not your friend**. Get out from behind it and center yourself to the group you are speaking to.

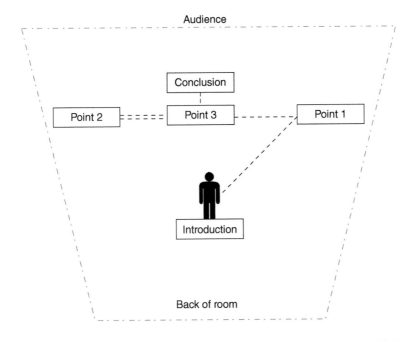

Having planned, purposeful movement helps engage audiences and create distinctions between your major ideas. This involves following the Speakers Diamond (see above figure). Start at the center of the audience. When you transition between your introduction and point 1, walk to one side of the room and stand in a spot. Repeat this process between points 1 and 2 where you walk to the opposite side. At point 3, return to the center of the room and take a step forward for your conclusion.

APPENDIX H: PRACTICE, FEEDBACK, AND ANXIETY REDUCTION TECHNIQUES FROM CHAPTER 9

Practice Step 1

Audio record a verbal run through. You may use notes by placing main points, details to remember including numbers, important connections, and sources in large print on 3×5 index cards. This is to keep you on track and is formally known as a presentation outline. Recording the verbal portion of your run-through provides you with three important pieces of feedback, (i) the amount of time it took from start to finish, (ii) the clarity and logical sequence of information, and (iii) the sound of your vocal delivery. Listen to the recording as an objectively as possible and evaluate yourself using the measures of each element as:

0 – Not Present   1 – Weak   2 – Good   3 – Strong

Audio self-evaluation

Time: _____

0 1 2 3   Attention getting device

0 1 2 3   Purpose statement

0 1 2 3   Established credibility

0 1 2 3   Previewed main points

0 1 2 3   Logical sequence of main ideas

0 1 2 3   Transition statements

0 1 2 3   Language is vivid/clear/fluid

0 1 2 3   Voice is used effectively

0 1 2 3   Conclusion is effective

0 1 2 3   Overall effectiveness

Comments:

_____

_____

_____

_____

_____

Practice Step 2

Video record a formal run-through using slides and presenting as you would to an actual audience. Once again, use bulleted note cards to keep you focused and break the habit of looking back at a projected slide. Place the cell phone or camera far enough back to capture you and your slides in the frame. Remember the tip from Chapter 5 on creating slides using dark backgrounds and white text or inversing the color on charts and graphs. This makes for easier viewing. Watch your recording and provide a self-critique making additional changes in content or delivery. Check the time and adjust if necessary. Evaluate elements of presentation accordingly:

0 – Not Present   1 – Weak   2 – Good   3 – Strong

Video Self-Critique:

Time:_____   Comments:

Content:

0 1 2 3   Gained audience's attention (AGD)

0 1 2 3   Introduced topic

0 1 2 3   Reinforced credibility

0 1 2 3   Previewed main points

0 1 2 3   Clear organization pattern

0 1 2 3   Included transition statements

0 1 2 3   Evidence of audience analysis

0 1 2 3   Strong conclusion

0 1 2 3   Final closing statement or Call to action

Delivery:

0 1 2 3   Gestures and nonverbal delivery

0 1 2 3   Use of voice and inflection

0 1 2 3   Effective use of slides

Practice Step 3 Peer-Evaluation

Time:_____   Comments:

Introduction
- ❏ Gained attention and interest
- ❏ Introduced topic clearly
- ❏ Motivated the audience to listen
- ❏ Established credibility
- ❏ Previewed remainder of speech

Discussion
- ❏ Main points well organized/developed
- ❏ Clearly focused topic
- ❏ Language: vivid, clear, creative
- ❏ Identified with audience
- ❏ Transitions: quality, points well identified

Supporting Materials
- ❏ Quality sources
- ❏ Solid evidence

Conclusion
- ❏ Reinforce purpose
- ❏ Summary clear
- ❏ Sense of closure

Delivery
- ❏ Level of animation/confidence/dynamism
- ❏ Gestures: effective, appropriate
- ❏ Voice clarity, vocal emphasis
- ❏ Extemporaneous style and use of notes

What questions might the audience ask?

_____

_____

Practice Step 4

Slide-critique.

_____

_____

APPENDIX I: TEN SESSION COMMUNICATION CURRICULUM

This curriculum that was created for the communication portion of The University of Alabama's Aeronautical Engineering REU (Research Experience for Undergraduates) program. REU is a National Science Foundation-supported program where undergraduates from across the country apply to hosting a university's field-specific 10-week research experience. The goal of REU is to spark students' interest in graduate school. The inclusion of a communication component in an REU grant proposal is a strong workforce readiness component of grant requests.

There are a variety of ways this curriculum could be incorporated, in part or in whole, into existing programs. The overall goals are to build communication competence and presentation skills alongside technical education.

Programmatic outcomes include communication skills training. There are 10, one and a half to two and a half-hour, sessions within the 10-week program that focus on communication training and preparing students to deliver research presentations. Topics covered are the goals of effective oral communication, nonverbal communication (body language, vocal variety, and visual aids), organization and structure of presentations, technical writing, and academic posters.

The objective of the communication portion of REU is to help students

- gain awareness of personal communication competence (strengths and weaknesses), learn goals of effective communication,
- learn to craft clear messages, construct effective visual aids, cultivate dynamic delivery, and
- calm public speaking anxiety through construction and delivery of a
  - Two to three-minute Poster Presentation;
  - Two to three-minute REU research video;
  - Five to eight-minute initial research presentation; and
  - Final ten to twelve-minute research presentation
- learn and apply technical writing skills and produce a research abstract for submission to a national conference.

Session 1: Oral Communication Seminar
Session 2: Presentation Workshop (Initial research plan presentation/feedback)
Session 3: How to Make a Research Video
Session 4: Research Plan Presentation for Faculty
Session 5: Making a Technical Poster Seminar
Session 6: Technical Writing Seminar
Session 7: Practice Research Presentations (updated presentation)
Session 8: Technical Poster / Abstract Workshop
Session 9: Poster Presentations and Video Reveal for Faculty
Session 10: Research Presentations for Faculty

Session 1: Initial Communication Seminar (2–2.5 hours)

The objective of this seminar is for students to:

- Articulate the relevance of oral and written communication skills to the field of engineering.
- Gain self-awareness of personal communication strengths and weaknesses.
- Understand the goals of effective communication.
- Learn basic presentation organization and delivery.

Pre-meeting Assignment

Instructors email students with initial assignment prior to first communication workshop. Students are to find current research or trade articles examining the importance of communication skills within the field of engineering. Each student is to upload the article citation into a shared Google Doc (or similar file sharing box). This eliminates redundancy and allows students to view each other's articles. *Students must prepare a three-minute summation of their article to present at the first workshop.

> Example e-mail:
>
> Hi All,
>
> For our first meeting, please find a scholarly journal or trade (engineering publication/ magazine) article that ties the importance of communication or speaking/presentation skills to the practice and profession of engineering. Place the citation on the Google Doc to make sure there are no duplicates and be ready to give us a three-minute summation at our first meeting.

Lesson Plan

- Article Summaries
  - Place students in small groups of three or four.
  - Group Instructions: Each student has three minutes to summarize their article.
  - Instructor acts as facilitator timing each three-minute round and moves group along.
  - After each member of the small group has presented, each group chooses one individual who had the most compelling article to present to the class.
  - Ask students about themes about the role of communication within engineering.
- Communication Self-Assessment
  - Handout self-assessment and have students rate themselves on the Likert-Scale for each statement. Likert scale ranges from 1 (Very Little) to 7 (Excellent) for each communication competence. Allow three to five minutes for completion.
  - After completion of Self-Assessment, pair students to share one or two perceived weaknesses (statements they scored 4 or below on) and two of their strengths with one another (statements they scored a 5 or above on). They are to explain why they believe each to be a weakness or strength.

- ○ Ask about common themes in perceived weaknesses/perceived strengths. Discuss importance of self-awareness in building communication skills and continued professional development.
- Teach goals of effective oral communication: (i) to create shared meaning/common ground, (ii) express understanding, and (iii) convey value and respect for audience.
- Tips for presentations and framework. (*Handout provided*)
  - ○ What do they NOT like about other's presentations? What makes a presentation bad?
  - ○ How do we make a good presentation? First Consider the 5W's and H.
  - ○ How to construct a presentation outline.

Assignment: Students are to prepare a five to eight-minute presentation of research topic, purpose of research, and methodology with visual aid. Presentation into: (i) purpose of research including questions/hypothesis, (ii) methodology, and (iii) expected results.

Session 1 Handout:

Communication Skills, Presentation Tips, and Techniques

Communication Assessment Questions

Assess your personal ability using the scale from 0 (none) to 7 (great) for each of the following as they apply to the work environment or professional situations.

0 – No Ability   7 – Great!

0-------1--------2-------3-------4-------5-------6-------7

1. Feeling confident to share your ideas verbally. _____
2. Communicate verbally with those in your team. _____
3. Communicate verbally to those in management or leadership positions. _____
4. Share ideas in a formal group setting (i.e. meeting). _____
5. Advocate for a specific action or point of view. _____
6. Understanding other's motivations/perspective. _____
7. Listening to gain understanding. _____
8. Reading someone's body language. _____
9. Awareness of the nonverbal (body language) cues you display during an interaction. _____
10. Self-awareness of vocal variety (volume, pitch, inflection) used during interactions. _____

In pairs, discuss your perceived communication competence weaknesses and strengths. What evidence or experiences led you to score yourself the way you did. Choose two communication skills you scored 4 or less on and explain why you feel these are areas of weakness for you.

_____

_____

Choose two communication skills you scored 5 or more on and explain why you feel these are areas of strength for you.

_____

_____

Goals of Communication

1. Create shared meaning.
2. Express understanding.
3. Convey value and respect.

Effective Presentations

> Too often I have been trapped in presentations that were painfully long, overly serious, and under organized. And I have been guilty of all of the above!
>
> *Unknown*

   *Examples of the Best and Worst presentations you have witnessed.

Level of Anxiety in Public Speaking Situations
You've been asked to speak to a group of professional peers or to deliver a presentation to potential clients/investors. Rate the level of anxiety you experience prior to speaking?

0 – No sweat!   10 – I'd rather be mauled by a bear!
    0------1--------2-------3-------4-------5-------6-------7------8------9------10

Guidelines for Creating Your Presentation:
Step 1: Consider the 5W's and H.

Who am I speaking to?
What am I trying to accomplish?
Where am I speaking?
When: time constraints.
Why is the information important? (Who Cares???)
How am I going to present it?

Step 2: Create an outline

Introduction

- Attention Getter
- Reveal Topic
- Objective of Research (why it matters?)
- Preview Main Points

Body:

1. Purpose of Study (Why is it important? Research Questions/Hypothesis)
2. Methodology
3. Results/Implications/Future Research

Conclusion:

- Review
- Strong Closing Statement (Return to AGD)

Step 3: Visual Aids

Crafting clean visual aids, AV or handouts. What are the presentation pitfalls?

Step 4: Practice

Now that you have your material well prepared, delivery is crucial.

Memorize introduction and conclusion.
PRACTICE! Run through entire presentation and TIME it.
Audio or video tape yourself.
Be aware of any verbal fillers (uh, um, and, like, ya know).
Arrive early and get a feel for the space and set-up.
Do not hide behind a podium.
SMILE AND MAKE EYE CONTACT.
Extemporaneous Style
Use pauses or silence strategically.
Be distraction-free.
Read the audience's body language – Are they engaged?
Enjoy it and they will too.
Strong finish: *Call to action!*

*Do not go over allotted time (rehearse).

Fielding Questions

Encourage them to ask questions (Do not be afraid of silence).

Repeat audience questions so the entire audience can appreciate your answer.

Keep eye contact with the person who posed the question.

Do not know? Say "I don't know the answer to that off hand, but I can find out and get back to you." Make sure to give me your e-mail address after the presentation.

Don't make promises you don't intend to keep....

Notes: _____

_____

Session 2: Presentation Workshop (2–2.5 hours)

The objective of this workshop is for students to:

- Develop improved public speaking skills through practice and feedback.
- Practice providing peers with constructive feedback in an encouraging manner.
- Create a plan for improvement through self-assessment.

Lesson Plan: Instructor will facilitate timing, encouraging and constructive group feedback, and recording of each student presentation. For the five to eight-minute presentation, time signals are shown at five, six, and eight-minute marks. Instructor will complete a short feedback form for each speaker. Recordings of each presentation should be uploaded to a shared file for students to view personal performances and complete self-critique.

Instructor Feedback

- Aspects of presentation you did well:
- Aspects of presentation that need improvement:

Self-Assessment

1. Was the research topic and significance clearly communicated to audience?
2. Material (clarity, organization, language)
   Strengths:
   Weaknesses:
3. Delivery (nonverbal gestures, confidence, voice)
   Strengths:
   Weaknesses:

4. What improvements will you implement for next time?

Assignment: Email students link to shared file with presentation videos and the self-assessment questions. They are to watch the video, complete the assessment, and email responses back to instructor.

Session 3: How to Make a Research Video (2–2.5 hours)

The objective of this seminar is for students to:

- Understand the necessary components of a research video
- Develop a short narrative of their research
- Employ creativity in video shoot
- Learn tips on filming – video editing

Lesson Plan: The instructor will show previous REU videos and allow group to provide feedback. Example from UA REU site: http://reu.eng.ua.edu/programs/fm-ace/past-projects

What did they like or not like about each video? Did they see elements that they would like to include in their videos?

Instructor will lead a discussion on keys to a successful video. Begin by asking students to write down ideas about the following:

- Construct a list of the most important ideas to communicate about your research.
- How could you explain the research to a generalizable audience in a compelling/fun manner?
- Think of yourself as a science video host, what would you need visually to help explain the research?

Points to remember:

- Have fun with it. Use the lab or equipment as props in the shoot, dress the part, think visually.
- Get equipped. Students may check out video equipment through the university's resource center/library.
- If using phone, shoot vertically, not horizontally.

End session discussing research plan presentations to faculty planned for the following session.

Assignment: Encourage them to take what they learned from feedback and work on any area of improvement as they practice. Students should upload their AV slides to a shared online file prior to presentations next session, allowing for easier transitions between speakers.

Session 4: Research Plan Presentation for Faculty (2–2.5 hours)

The objective of this seminar I for students to:

- Present initial research plan presentations for faculty advisors
- Gain insight from advisors on communicating research

Instructor facilitates presentation session by timing each speaker, facilitating feedback from advisors, and moving session along. Instructor may choose to record presentations for students to review.

Assignment: NA

Session 5: Making a Technical Poster Seminar (1.5–2 hours)

The objective of this seminar is for students:

- Learn about conference poster sessions
- Understand purpose of posters as visual representations of project
- Learn how to construct a technical poster

Lesson Plan: As these students have not been exposed to academic or technical conferences, the instructor will discuss the conference setting, goals of individuals attending and representing work at conferences, and describe a poster session setting. Poster sessions include many presenters showcasing their work to passersby. Those attending a poster session are a voluntary audience in that they are interested in the topics and work being presented. Presenters will have prepared a visual and narrative display of their research beginning with the purpose of the research, questions or hypothesis, methods, results, and implications. The poster also credits the institution and research contributors. Presenters should be able to orally describe the research in two minutes, peaking the interest of the listener that they may ask more detailed questions of the presenter.

Print out, on legal-sized paper, a poster template to distribute to each student. Describe each section of the poster and how, like reading, the arc of the research should begin top left and end bottom right. Poster creation tips:

- What? How they are used. Discuss conference set-up and poster session objectives, layout, and the 3MT
- Templates – Found online or in slide templates

General guidelines include:

- Easy to read sub-headings for each section – 36 pt. font
- Captions and labels should be 18 pt. font

- Approximately 100 words per section – 300–800 total word count with 24 pt. font
- Use bullets or numbering for easy reading
- Use black and primary colors
- Do not have too many data points or labels on charts or graphs
- Import images as jpegs with high quality resolution

Design-wise you are aiming for a passerby to be able to read slide headlines from 10 feet and understand the topic, motivation, and important findings from a social distance of three to five feet and within the first 20 seconds of looking at the poster. Choose a sans serif font as they are more legible and easier to read from a distance. Use solid backgrounds that are dark with black for text and primary colors for labeling and effects. With dark backgrounds, white text boxes do not need a border. With lighter backgrounds use borders around text boxes to define the section. Keep some white space between the text and border of each section's box.

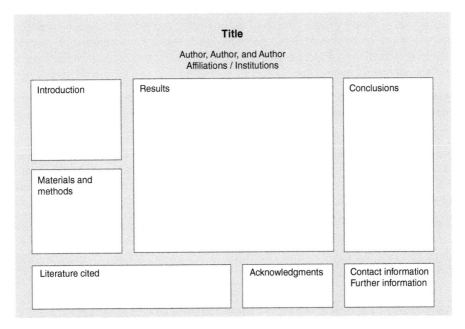

Assignment: Begin working on poster.

Session 6: Technical Writing Seminar – Begin Working on Abstracts (1.5–2 hours)

The objective of this seminar is for students to:

- Understand the goals and challenges of technical writing
- Examine purpose and elements of an abstract

Lesson Plan: Print out various field specific journal article abstracts. Try and choose some that you believe are good, clear examples that draw the reader in and some poor examples. Make two copies of each example abstract. Pass them to students, giving two students the same example. Students read the abstract and partner with the individual who read the same example. Ask students to discuss the study's purpose and findings with their partner and list good or bad elements of the writing example.

Discuss goals of technical writing: (i) Convey information in an effective manner. (ii) Make complex concepts clear and easy to understand. (iii) Communicate a process simply and logically. Technical writing when applied to an abstract, needs to consist of one or more well-developed paragraphs that act as a summation of their research, written for a wider audience. Abstracts will be used to place on the posters and for conference submissions.

Activity Instructions to Students: Ask students to think of one piece of lab equipment that they are learning how to use. How clear were the instructions? Now, it is their turn.

Instructions: Take out your phone and turn it off. Now write directions to guide someone who has never used a cell phone to turn it back on and make a phone call. You may re-activate your phone and simulate the process. You may not use diagrams or visuals, only words, and you must use complete sentences. Aim for brevity.

Assignments: Begin to construct a draft of their abstract. Update and practice full research presentation, adding newer work and findings. Students will deliver a run-through during the next session. They need to bring/send updated slides. They will have up to 10 minutes to present.

Session 7: Practice Research Presentations (2–2.5 hours)

The objective of this seminar is for students to:

- Practice research presentation
- Develop speaking and presenting skills
- Decrease anxiety associated with public speaking
- Receive further feedback

Lesson Plan: Instructor facilitates session by timing each speaker, allowing peers to offer feedback, and providing further positive/constructive feedback on speaking skills and visual aid slides. Instructor may choose to record presentations for students to further review. Evaluation sheet from Appendix F may be utilized.

In preparation of the poster presentation, watch examples of 3MT (Three Minute Thesis) speakers. Videos are easily found online. Ask students how graduate students made their topics appealing by orally summarizing their study in three minutes or less.

Assignment: Students prepare a rough draft of their poster, including the draft of the abstract, on a AV slide template and a two to three-minute research summary to present at the next session. Students should bring their laptops with the poster template and abstract to the following workshop.

Session 8: Technical Poster/Abstract Workshop (2–2.5 hours)

The objective of this seminar is for students to:

• Practice delivering a poster presentation.
• Receive feedback on design elements of poster.
• Develop technical writing skills.

Lesson Plan: Each student projects poster slide on AV screen and presents their two to three-minute summary of research. Following each speaker, allow peers to provide feedback about content and design of the slide. Make sure that this is a beneficial process, starting with the positive elements of the poster, asking questions about design choices, and providing suggestions.

After all poster presentations are complete, ask students to pull up their abstracts on their laptops and partner with someone to read one another's abstract and give helpful feedback with the aim being clarity of language and process explanations.

Assignment: Make any necessary edits/changes to poster, practice presentation, and complete research video for faculty advisor seminar the following week. Students must send their poster slide and video link or file to a shared drive to be pulled up at the beginning of the session next week.

Session 9: Poster Presentations and Video Reveal for Faculty (2–2.5 hours)

The objective of this seminar is for students to:

• Increase speaking and presentation proficiency.
• Gain confidence in presenting to faculty.

Lesson Plan: Pull up shared drive prior to the session and make sure all AV, including sound are working. Instructor facilitates each student presenting their poster presentation, using AV to project slide, and showing the video to the faculty. Allow faculty to provide positive feedback or constructive suggestions to each speaker.

Assignment: Practice their final research presentation, 10–12-minutes, to deliver to faculty at last meeting. Students should update slide in the shared drive.

Session 10: Research Presentations for Faculty (2.5–3 hours)

The objective of this seminar is for students to:
• Deliver a final research presentation with confidence to faculty.

Lesson Plan: Instructor facilitates presentations, records time and give time signals at 10, 11, and 12 minutes (signaling to wrap up if student goes over 12). Record presentations and upload to a Drive folder that can be shared with students and REU faculty.

Example REU Schedule: Aeronautical Engineering REU Site: Fluid Mechanics with Analysis using Computations and Experiments (FM-ACE), NSF Grant Award Number: 1358991.

Week	Monday	Tuesday	Wednesday	Thursday	Friday	Weekend
1	Meet with mentor Scheduled times between 9 and 11 a.m. 12–1 p.m. ORIENTATION LUNCH RESEARCH	RESEARCH	RESEARCH RESEARCH	RESEARCH	RESEARCH Seminar lunch Fundamentals of Research seminar 10:30 a.m. – 12:30 p.m. Lab Tour 5:30 p.m. Welcome Dinner	
2	Memorial Day	RESEARCH 11 a.m. – 1 p.m. LUNCH Meet with program evaluators/ Presentation skills video	Literature seminar 9–11 a.m. Rodgers Engineering Library Scholars station lab	**Session 1: Oral Communication Seminar 9–11 a.m.**	RESEARCH	SATURDAY Day trip to Huntsville Space and Rocket Center
3	MEET WITH ADVISOR RESEARCH	Flow Measurement seminar 9–11 a.m. Lab Tour RESEARCH	RESEARCH RESEARCH	RESEARCH **Session 2: Presentation Workshop 9–11 a.m.** RESEARCH	RESEARCH	

(Continued)

Week	Monday	Tuesday	Wednesday	Thursday	Friday	Weekend
4	**Session 3: How to Make a Research Video 9–11 a.m.**	RESEARCH **Session 4: Research Plan Presentation for Faculty 10 a.m.–12 p.m.** Lab Tour	RESEARCH	RESEARCH Professional development Workshop 11 a.m. – 1 p.m. <u>Lunch provided</u> Lab Tour	RESEARCH	
5	RESEARCH MEET WITH ADVISOR	RESEARCH RESEARCH	RESEARCH	RESEARCH RESEARCH **Session 5: Making a Technical Poster Seminar 10 a.m.–12 p.m.** Lab Tour RESEARCH		FIELD TRIP
6	RESEARCH RESEARCH	**Session 6: Technical Writing Seminar 9–11 a.m.** RESEARCH	RESEARCH	RESEARCH <u>Seminar lunch</u> How to apply for graduate school and taking the GRE 10:30 a.m. – 12:30 p.m. Lab Tour RESEARCH	RESEARCH	

7	RESEARCH	RESEARCH	4 July Holiday (university closed)	RESEARCH	RESEARCH **Session 7: Practice Research Presentations 10:30 a.m. – 12:30 p.m.** FIELD TRIP
8	Visit to GA Tech	RESEARCH	RESEARCH	RESEARCH **Session 8: Technical Poster/Abstract Workshop Lunch 11:30 a.m. – 1:30 p.m.** RESEARCH	RESEARCH
9	RESEARCH	RESEARCH	RESEARCH	RESEARCH	RESEARCH
10	**Session 9: Poster Presentations and Video Reveal for Faculty** 10 a.m. – 2 p.m. <u>Lunch</u> Discuss APS abstract submission	RESEARCH	RESEARCH	RESEARCH **Session 10: Research Presentations for Faculty** Breakfast and lunch with all REU site contributors/students 8:30 a.m. – 2 p.m.	RESEARCH SUBMIT FINAL WRITTEN REPORT TO ADVISOR Students leave for home

INDEX

Engineered to Speak: Helping You Create and Deliver Engaging Technical Presentations, First Edition.
Alexa S. Chilcutt and Adam J. Brooks.
© 2019 by The Institute of Electrical and Electronics Engineers, Inc. Published 2019 by John Wiley & Sons, Inc.

Books in the
IEEE PCS PROFESSIONAL ENGINEERING COMMUNICATION SERIES

Sponsored by IEEE Professional Communication Society

Series Editor: Ryan Boettger

This series from IEEE's Professional Communication Society addresses professional communication elements, techniques, concerns, and issues. Created for engineers, technicians, academic administration/faculty, students, and technical communicators in related industries, this series meets the need for a targeted set of materials that focus on very real, daily, on-site communication requirements. Using examples and expertise gleaned from engineers and their colleagues, this series aims to produce practical resources for today's professionals and pre-professionals.

Information Overload: An International Challenge for Professional Engineers and Technical Communicators · Judith B. Strother, Jan M. Ulijn, and Zohra Fazal

Negotiating Cultural Encounters: Narrating Intercultural Engineering and Technical Communication · Han Yu and Gerald Savage

Slide Rules: Design, Build, and Archive Presentations in the Engineering and Technical Fields · Traci Nathans-Kelly and Christine G. Nicometo

A Scientific Approach to Writing for Engineers and Scientists · Robert E. Berger

Engineer Your Own Success: 7 Key Elements to Creating an Extraordinary Engineering Career · Anthony Fasano

International Virtual Teams: Engineering Global Success · Pam Estes Brewer

Communication Practices in Engineering, Manufacturing, and Research for Food and Water Safety · David Wright

Teaching and Training for Global Engineering: Perspectives on Culture and Professional Communication Practices · Kirk St.Amant and Madelyn Flammia

The Fully Integrated Engineer: Combining Technical Ability and Leadership Prowess · Steven T. Cerri

Culture and Crisis Communication: Transboundary Cases from Nonwestern Perspectives · Amiso M. George and Kwamena Kwansah-Aidoo

Engineered to Speak: Helping You Create and Deliver Engaging Technical Presentations, First Edition.
Alexa S. Chilcutt and Adam J. Brooks.
© 2019 by The Institute of Electrical and Electronics Engineers, Inc. Published 2019 by John Wiley & Sons, Inc.